Advances in Prolactin

Progress in Reproductive Biology

Vol. 6

Series Editor: P.O. Hubinont, Brussels
Assistant Editor: M. L'Hermite, Brussels

Editorial Board: *E. Beaulieu*, Paris; *E. Diczfalusy*, Stockholm; *K. Eik-Nes*, Trondheim; *J. Ferin*, Louvain; *P. Franchimont*, Liège; *F. Hefnawi*, Cairo; *J.J. Keller*, Zürich; *B. Lunenfeld*, Tel-Aviv; *L. Martini*, Milano; *J. Mulnard*, Brussels; *F. Neumann*, Berlin; *M. Neves e Castro*, Oss; *H. Pedersen*, Copenhagen; *J.C. Porter*, Dallas, Tex.; *P. Pujol-Amat*, Barcelona; *R.J. Reiter*, San Antonio, Tex.; *P. Soupart*, Nashville, Tenn.; *H.-D. Taubert*, Frankfurt am Main; *M. Thiery*, Gent

S. Karger · Basel · München · Paris · London · New York · Sydney

Proceedings of the Satellite Symposium to the 6th International Congress of
Endocrinology, Adelaine, February 18-19, 1980

Advances in Prolactin

Volume Editors
M. L'Hermite, Brussels
S.J. Judd, Bedford Park

76 figures and 23 tables, 1980

QP251
A1
P76
v.6
1980

S. Karger · Basel · München · Paris · London · New York · Sydney

Progress in Reproductive Biology

Vol. 5: Seasonal Reproduction in Higher Vertebrates. Symposium on the Seasonal Reproduction in Higher Vertebrates, Richmond, Va., 1978.
Reiter, R.J., San Antonio, Tex. and Follett, B.K., Bristol (eds.)
VI + 222 p., 84 fig., 7 tab., 1980. ISBN 3-8055-0246-X

Vol. 7: Blastocyst-Endometrium Relationships. 7th Seminar on Reproductive Physiology and Sexual Endocrinology, Brussels, May 21–24, 1980.
Leroy, F., Brussels, Finn, C.A., Liverpool, Psychoyos, A., Paris, and Hubinont, P.O., Brussels (eds.)
X + 338 p., 94 fig., 30 tab., 1980. ISBN 3-8055-0988-X

All rights reserved.
No part of this publication may be translated into other languages, reproduced or utilized in any form or by any means, electronic or mechanical, including photocopying, recording, microcopying, or by any information storage and retrieval system, without permission in writing from the publisher.

© Copyright 1980 by S. Karger AG, P.O. Box, CH-4009 Basel (Switzerland)
Printed in Switzerland by Tanner & Bosshardt AG
ISBN 3-8055-0859-X

Contents

Foreword VIII

Lactotrope: Stimulation and Regulation of Its Activity

Thorner, M.O. and MacLeod, R.M. (Charlottesville, Va.): The Lactotrope – Regulation of Its Activity... 1
    Discussion by Herschman, Eastman, Wuttke and Jacobs 22
Adler, R.A.; Brown, S.J., and Sokol, H.W. (Hanover, N.H.): Characteristics of Prolactin Secretion from Anterior Pituitary Implants 24
    Discussion by Bern, Neill, Leong, Herschman, Wuttke, Greenwood and Kelly 30
Dufy, B.; Vincent, J.-D.; Gourdji, D., and Tixier-Vidal, A. (Bordeaux): Electrophysiological Study of Prolactin-Secreting Pituitary Cells in Culture 31
    Discussion by Friesen, Bern, Neill, Thorner and Willoughby 42
Shin, S.H. and Chi, H.J. (Kingston, Ont.): Evidence for the Existence of the Physiological Prolactin-Releasing Factor and its Possible Role 44
    Discussion by Voogt and Neill 53
Kato, Y.; Matsushita, N.; Katakami, H., and Imura, H. (Kyoto): Effect of a Synthetic Enkephalin Analogue on Prolactin Secretion in Man and Rats.... 54
    Discussion by Veldhuis 60
Désir, D.; L'Hermite, M.; Van Cauter, E.; Caufriez, A.; Jadot, C.; Spire, J.-P.; Noël, P.; Refetoff, S.; Copinschi, G., and Robyn, C. (Brussels): Nyctohemeral Profiles of Plasma Prolactin after Jet Lag in Normal Man 61
    Discussion by Jacobs and Frantz 65
Buckman, M.T.; Srivastava, L.S., and Peake, G.T. (Albuquerque, N. Mex.): Regulation of Prolactin Secretion by Endogenous Sex Steroids in Man 66
Whitaker, M.D.; Corenblum, B.; Taylor, P.J., and Harasym, P.H. (Alberta): Control of the Hypoglycemia Release of Prolactin 77
Herman, V.S. and Kalk, W.J. (Johannesburg): Neurogenic Prolactin Release: Effects of Mastectomy and Thoracotomy 83
    Discussion by Adler, Bern and Buttle 86

Contents VI

Judd, S.J. (Bedford Park): Autoregulation of Prolactin Secretion 87
General Discussion. Chairmen: *Neill, J.* and *Judd, S.J.* 92
    Discussion by *Buckman, Thorner, Robyn, Kelly, Jacobs, Shin, Wuttke, Frantz, Voogt* and *Greenwood* 96

Prolactin Target Cell and Receptor

Shiu, R.P.C. (Winnipeg, Man.): The Prolactin Target Cell and Receptor. An Overview 97
    Discussion by *Guyda, Waters, Bohnet, Bern, Ahren, Carter, Barkey* and *Jacobs* 111
Barkey, R.J.; Lahav, M.; Shani, J.; Amit, T,; Youdim, M.G.H., and *Barzlai, D.* (Haifa): Induction of Hepatic Prolactin-Binding Sites in the Male Rat Liver: Role of Prolactin and Prostaglandins 113
Kelly, P.A.; Djiane, J., and *De Léan, A.* (Québec): Prolactin-Receptor Dissociation and Down-Regulation 124
    Discussion by *Herington* and *Waters* 135
General Discussion. Chairmen: *Guyda, H.* and *Falconer, I.R.* 137
    Discussion by *Eastman, Kelly, Barkey, Greenwood, Shin, Bern* and *Rillema* 137

Prolactin: Questions without Answers

Willoughby, J.O. (Bedford Park): Prolactin: Questions Without Answers 142
    Discussion by *Thorner, Bern, Neill, Buttle* and *Lyons* 164
Bern, H.A.; Loretz, C.A., and *Bisbee, C.A.* (Berkeley, Calif.): Prolactin and Transport in Fishes and Mammals 166
    Discussion by *Falconer* and *Greenwood* 171
Bohnet, H.G.; Gabel, A.K., and *Kreutzer, P.* (Münster): Prolactin Secretion Patterns in Patients with Mammary Tumors. Preliminary Results 172
    Discussion by *Judd* and *Eastman* 178

Clinical Hyperprolactinaemia: Its Effects and Treatment

Gross, B.A.; Haynes, S.P.; Eastman, C.J.; Balderrama-Guzman, V., and *del Casillo, V.* (Canberra): A Cross-Cultural Comparison of Prolactin Secretion in Long-Term Lactation 179
    Discussion by *Robyn, Eastman* and *Judd* 186
Coutts, J.R.T.; Fleming, R.; Craig, A.; Barlow, D.; England, P., and *Macnaughton, M.C.* (Glasgow): Role of Transient Hyperprolactinaemia in the Short Luteal Phase 187
    Discussion by *Judd, Haines* and *Robyn* 193
Nillius, S.J. (Uppsala): Medical Therapy of Prolactin-Secreting Pituitary Tumors 194
    Discussion by *Friesen, Thorner, Frantz, Dolman, Carter, Cowden, Judd* and *Bohnet*... 218

Contents

Scanlon, M.F.; Rodriguez-Arnao.; McGregor, A.M.; Pourmand, M.; Weightman, D.R.; Cook, D.B.; Gomez-Pan, A.; Lewis, M., and *Hall, R.* (Newcastle): A New Approach to the Identification of Prolactin-Secreting Microadenomas? . 222
  Discussion by *Adler, Eastman, Harding, Cowden* and *Hershman* 230
Hall, K.; McGregor, A.M.; Scanlon, M.F., and *Hall, R.* (Newcastle): Metrizamide Cisternography in the Assessment of Pituitary Tumor Size, with Special Reference to the Demonstration of Tumour (Prolactinoma) Shrinkage on Bromocriptine Therapy 238
  Discussion by *McGregor, Jacobs, Thorner, Buckman, Frantz* and *del Pozo* 242
Dolman, L.I. (San Francisco, Calif.): Primary Hypothyroidism: Abnormalities of Prolactin Secretion and Pituitary Function 244
  Discussion by *Adler, Frantz, Hershman* and *Jacobs* 251
del Pozo, E. and *Falaschi, P.* (Rome): Prolactin and Cyclicity in Polycystic Ovary Syndrome 252
  Discussion by *Adler, Beumont, Ajayi* and *Jacobs* 258
Evans, L.E.J.; Beumont, P.J.V.; Luttrell, B.; Henniker, A.J., and *Penny, J.* (Sydney): Prolactin Response as a Guide to Neuroleptic Treatment of Schizophrenia 260
General Discussion. Chairmen: *Frantz, A.G.* and *Eastman, C.* 264
  Discussion by *Molinatti, Judd, L'Hermite, Dolman, Cowden, del Pozo, Isaac, Thorner* and *Ajayi* 264

Foreword

The *Prolactin – 1980 and Beyond* workshop was held in Adelaide, South Australia, as a satellite meeting of the Sixth International Congress of Endocrinology.

This workshop provided an ideal opportunity to gather together comparative endocrinologists, molecular biologists, neurophysiologists and clinicians to discuss new aspects of their common interest – prolactin. The workshop consisted of overviews of important areas, short communications and free discussion. The discussion was wide-ranging and uninhibited and the presentations were of high quality, reflecting not only the current state of the art, but also the promise of the future. The success of the workshop was ensured by the quality of the investigators attracted to the meeting and the ability of the chairmen to allow freedom of discussion whilst maintaining an orderly framework.

This publication aims to bring as much as possible of the proceedings of this workshop to a wider group of basic scientists and clinicians. In doing so, we hope to have communicated some of the excitement we shared at this meeting.

Finally, we would like to acknowledge the support of Sandoz Limited and the Postgraduate Foundation in Medicine of the University of Adelaide for their help in funding this meeting and in the edition of its proceedings.

Stephen J. Judd,
Flinders Medical Centre,
Flinders University, Adelaide

Marc L'Hermite
Hôpital Universitaire Brugmann,
Université Libre de Bruxelles, Bruxelles

Lactotrope: Stimulation and Regulation of Its Activity

The Lactotrope — Regulation of Its Activity

M.O. Thorner and R.M. MacLeod

Department of Internal Medicine, Division of Endocrinology and Metabolism, University of Virginia School of Medicine, Charlottesville, Va.

The identification of prolactin as a distinct and separate hormone from growth hormone only 9 years ago opened a new era in neuroendocrinology. Rapidly, the small quantities of purified hormone available were utilized to develop a specific and sensitive radioimmunoassay which enabled the study of the physiology of human prolactin to proceed rapidly. The physiology of prolactin has been comprehensively reviewed [19, 32, 52]. In this paper we will briefly summarize some aspects of the control of prolactin secretion and then turn our attention to some *in vitro* and *in vivo* studies on prolactin secretion which we hope will be relevant to the better understanding of the normal control of prolactin and also the pathophysiology of hyperprolactinemia and its management by medical means.

Patterns of Prolactin Secretion in Man

The changes in circulating levels of prolactin from fetal life to old age are shown diagrammatically in figure 1. In the fetus up to 25 weeks of gestation circulating prolactin levels are slightly higher than in normal adults. From 25 weeks until birth prolactin levels in the fetus rise progressively to levels which may exceed 500 ng/ml. After birth there is a fall over the first week to between 50 and 80 ng/ml and at about 6 weeks levels have fallen to pre-pubertal levels of 15 ng/ml. Levels are maintained at this level until at puberty there is a small rise in prolactin levels at stage 2 of breast development in girls [53]. Serum prolactin levels in girls after menarche are higher than before puberty and than in boys or men although there is overlap in the ranges.

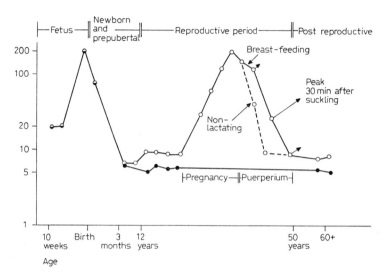

Fig. 1. Average serum prolactin levels at various periods of life in human male (●) and female (○) subjects. The length of the arrows in females who are breast-feeding is intended to indicate the magnitude of the increase in serum prolactin after each episode of suckling. [Reproduced from ref. 19 with permission.]

During pregnancy prolactin levels rise. The highest levels seen are in the third trimester. Following delivery prolactin levels fall and are normal at 2 weeks if lactation does not take place. However, if breast-feeding occurs the elevated prolactin levels are maintained and with suckling prolactin levels rise rapidly over 30 min and fall gradually until the next feed. As the interval from delivery increases in the lactating woman, the basal prolactin levels fall and the response to suckling declines. By 3 months, in western society, basal prolactin levels are normal, and there may be no rise with suckling [42].

Prolactin levels do not remain constant throughout the 24-hour period. Prolactin levels rise shortly after falling asleep and then fall during the morning. The rise in prolactin is related to sleep rather than to time of day, since shifts in prolactin secretion occur with acute day night sleep reversal, with daytime naps and are delayed with delays in onset of sleep. Prolactin rhythms are independent of those of the other anterior pituitary hormones. Suppression of the rhythm occurs with exogenous glucocorticoids and Cushing's syndrome [12, 26]. The surges in prolactin release occur immediately following REM sleep, i.e. during non-REM sleep. Since non-REM

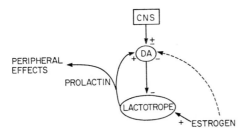

Fig. 2. Schematic representation of the interrelationships between the lactotrope dopamine and estrogen in the control of prolactin secretion.

sleep is associated with activation of serotonergic mechanisms this neurotransmitter has been implicated in these prolactin surges.

In the rat there is a clear proestrus surge of prolactin secretion that occurs on the afternoon of proestrus. Because of this, several investigators looked for changes in prolactin levels during the menstrual cycle. Initially it was reported that there was a midcycle increase in prolactin levels with a secondary rise during the luteal phase of the cycle. Some subsequent studies have failed to observe these changes [for review see 52]. It thus appears that the pattern of prolactin secretion during the female reproductive cycle is different in the human and the rat.

Prolactin is secreted in a pulsatile manner although there is no fixed periodicity. Prolactin levels rise in response to physical and emotional stress and prolactin levels also rise reflexly with suckling. Prolactin levels may also rise in normal women with stimulation of the nipple by a sexual partner, but not by self-stimulation, and during intercourse.

From the above brief summary of the pattern of prolactin secretion observed in the human it is clear that there are not only minute-to-minute changes in the secretory pattern, but also longer cycles of changes of release. Thus, for example, during pregnancy there is a progressive rise in prolactin levels in both fetus and mother. The rise in the prolactin levels in the fetus is independent of the presence of an intact hypothalamus. i.e. in anencephaly [3]. Thus, the prolactin rise in the fetus probably reflects changing estrogen levels which become extremely high in both fetus and mother during pregnancy.

In figure 2 a simplified scheme by which some factors thought to effect prolactin release is shown. As will be further discussed below, estrogen probably acts directly at the pituitary to stimulate not only synthesis and secretion of prolactin, but also mitosis of lactotroph cells.

The Control of Prolactin Secretion

Prolactin is unique among the anterior pituitary cells in that it is under predominant tonic hypothalamic inhibitory control. Thus, if the connection between the hypothalamus and the pituitary is disrupted, then secretion of all the anterior pituitary hormones will become deficient with the exception of prolactin which will become excessive [55]. In this paper it is not intended to discuss the nature of the various putative prolactin-releasing factors (PRF) or prolactin release-inhibiting factors (PIF). It is clear that thyrotropin-releasing hormone (TRH) is capable of stimulating prolactin *in vivo* and *in vitro*. Whether TRH is the physiological PRF is unresolved since there is dissociation of prolactin and thyroid-stimulating hormone (TSH) secretion *in vivo* (e.g. during suckling) [20]. The nature of PIF, however, is almost certainly the catecholamine dopamine. It is possible that there are other PRF and PIF but their presence and physiologic importance remain to be demonstrated. Increasing evidence suggests that the majority of the minute-to-minute control of dynamics of prolactin secretion may be largely explained by changes in dopamine release from the hypothalamus. Studies in the rat have shown the expected changes in dopamine concentration in the portal capillaries to at least partially explain the rise of prolactin following cervical stimulation [15] and the suppression of prolactin release from the *in situ* pituitary seen in rats bearing an experimental rat pituitary tumor [13].

The tuberoinfundibular dopamine (TIDA) neurons secrete dopamine into the portal capillaries in the median eminence. The dopamine is then transported by the portal capillaries to the anterior pituitary to inhibit prolactin release. The feedback for prolactin release appears to be via a short loop. The tuberoinfundibular neurons are unique in that their dopamine turnover is directly related to prolactin, while dopamine turnover in other dopamine neurons is independent of prolactin [44]. Further dopamine receptor-blocking drugs are only capable of stimulating dopamine turnover in the TIDA neurons if the pituitary is intact; in hypophysectomized animals the drugs are ineffective in modifying dopamine turnover because of the absence of prolactin [2, 14]. Thus, the overall scheme for control of prolactin release appears very simple: prolactin is tonically secreted by the pituitary while its release is inhibited by dopamine secreted by the TIDA neurons (fig. 2). Their activity in turn is controlled by the secretion of prolactin. Very little is known about the intra-hypothalamic control of the TIDA neurons. However, presumably projections from the spinal cord and higher centers terminate on them to modulate dopamine release to control the

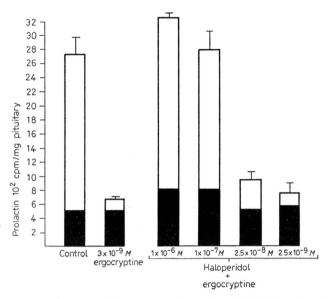

Fig. 3. Blockade of the ergocryptine-mediated inhibition of prolactin release by haloperidol. 3×10^{-9} M ergocryptine was incubated alone and in the presence of varying concentrations of haloperidol in flasks containing four hemipituitary glands; three flasks per group. [Reproduced from ref. 36 with permission.]

pattern of prolactin release; for example, during suckling or in response to stress. It may be anticipated that the firing of the TIDA neurons would be inhibited to allow prolactin release to be increased.

In vitro *Studies Using Rat Hemipituitary Incubations and the Pituitary Cell Column*

Dopaminergic Inhibition of Prolactin Release

Using *in vitro* incubations of hemipituitaries, *MacLeod* [31] and *MacLeod and Lehmeyer* [36] demonstrated that prolactin release could be inhibited by catecholamines including dopamine. Dopamine was the most potent inhibitor of prolactin release when compared to norepinephrine [34]. Subsequently the inhibitory effects of ergot derivatives, which were used as pharmacologic tools to inhibit prolactin release first in the laboratory and later in clinical practice, were found to be mediated through the same mechanism as dopamine. This is illustrated in figure 3 where the inhibitory effects

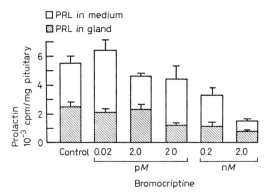

Fig. 4. Dose-related inhibition of prolactin release by bromocriptine. Four hemipituitary glands were incubated per flask in medium 199 alone or with increasing concentrations of bromocriptine; three flasks per group.

of ergocryptine are overcome, in a dose-dependent fashion by the dopamine antagonist, haloperidol [36]. The effect of different doses of bromocriptine on the pituitary content and release of newly synthesized prolactin is shown in figure 4. It is of interest that the α-adrenergic antagonists phentolamine and phenoxybenzamine are poor inhibitors of the effects of dopamine at the pituitary and thus these receptors are not α-receptors but dopamine receptors [9].

Dopamine receptors have been identified at the anterior pituitary using *in vitro* binding studies [8, 9, 14]. There appears to be two classes of binding sites for dopamine – higher affinity and lower affinity sites. The high affinity site is probably the biologically active one and has a dissociation constant (K_d) of 4.4×10^{-10} M. The dopamine agonists and antagonists compete competitively in these *in vitro* receptor assay systems. There appears to be fairly good agreement between the biological potency and the competitive binding potency of these compounds.

Estrogenic Stimulation of Prolactin Biosynthesis

Estrogens exert an extremely important role in regulation of the physiological prolactin production. Pituitary glands from female rats contain severalfold more prolactin than found in glands from male rats. Gonadectomy decreases pituitary prolactin content and biosynthetic capacity, both of which are restored by the administration of estrogens [35]. Specific estrogen receptors have been identified in pituitary tissue which are directly

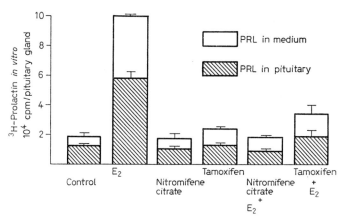

Fig. 5. Effect of *in vivo* estradiol and anti-estrogen administration on the *in vitro* synthesis and release of ³H-prolactin by the anterior pituitary.

stimulatory for the generation of messenger RNA which codes for prolactin synthesis [47, 48, 56]. Estrogen administration increases the circulating concentration of prolactin, but the primary action of the steroid is on the synthesis of prolactin. The stimulatory effect of estradiol on prolactin secretion may be mediated through modulation of the inhibiting action produced by dopamine. There is little doubt that dopamine-mediated inhibition of prolactin secretion is much decreased in pituitary glands following treatment with estradiol [27, 34]. Evidence is accumulating that the binding of ³H-estradiol to pituitary receptors is decreased following the administration of ergot alkaloids, e.g. ergocryptine, and may represent a mechanism through which the monoaminergic and estrogenic systems interact to regulate prolactin synthesis and release [39]. It is also probable that estradiol has an indirect action to stimulate prolactin secretion. Administration of the steroid has been shown to decrease the turnover of hypothalamic dopamine which subsequently decreases the inhibitory tone exerted on the pituitary, thus increasing prolactin secretion [18].

The role of anti-estrogens in the regulation of prolactin synthesis and release has recently been investigated. These drugs block the proestrus surge of prolactin [22, 41], but they do not alter basal serum levels even after chronic therapy [22, 24, 25]. We have studied the interaction of injected estradiol and anti-estrogens on the *in vitro* synthesis of prolactin [38]. The data in figure 5 indicate that the administration of estradiol to male rats caused a large increase in the *in vitro* synthesis of prolactin. In contrast, neither of the anti-estrogens, nitromifene citrate or tamoxifen, changed the basal rate of prolactin synthesis or release. Co-administration of estradiol and the anti-estrogens reduced the stimulatory effect of estradiol on prolactin synthesis. Thus, although anti-estrogens have little activity to alter basal prolactin synthesis and release, the drugs significantly reduce the stimulatory action of estradiol to increase prolactin synthesis. No effects of the anti-estrogen were observed on prolactin secretion.

The Intrinsic Tonic Nature of Prolactin Secretion: The Role of Calcium

When lactotroph cells are isolated from hypothalamic influences prolactin release becomes maximal. The stimulus for this secretion appears to be intrinsic to the cell and is poorly understood. The observation of action potentials in anterior (10% of cells) pituitary cells has led to the suggestion that these spikes, which have the characteristics of slow calcium spikes – i.e. being dependent on extracellular calcium rather than sodium ions – may participate in stimulus secretion coupling of prolactin, particularly since two factors known to effect prolactin release effect the spiking: TRH and dopamine, respectively, stimulate or inhibit spiking [50, 51]. Although the concept of stimulus secretion coupling has been acknowledged in studies on many secretory systems, including some cells of the anterior pituitary, its particular application to studies on the control of prolactin release by the lactotroph has curiously been largely neglected. We have performed some studies to investigate the effects on prolactin release of some factors which have been shown to alter stimulus secretion coupling and/or spiking of anterior pituitary cells.

The majority of these studies were performed using the perfused pituitary cell column of *Lowry* [30] and *Yeo et al.* [58]. It is probably the simplest and one of the most sensitive methods for examining control of pituitary hormone secretion and is thus admirably suited to investigating these questions. The system eliminates the hypothalamic input to the pituitary and is probably the closest model to an isolated perfused pituitary preparation; it allows administration of a test substance of known concentrations into the equivalent of portal capillaries and allows the effects to be directly monitored by hormone measurement in the effluent from the column. In this system the problem of diffusion into the tissue is obviated since the dispersed cells are isolated and supported by the inert matrix of Biogel.

Taraskevich and Douglas [50, 51], who recently described the presence of action potentials in about 10% of isolated anterior pituitary cells, proposed that these spikes might be a key process which modulates calcium

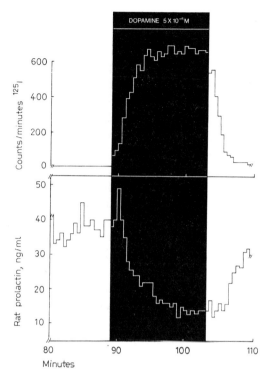

Fig. 6. The rapid onset and recovery from a 15-min dose of 5 µM dopamine on the release of prolactin from pituitary cell column. Upper panel: counts per minute 125I-luteinizing hormone used as indicator for dopamine concentration in 0.5-min fractions of eluate from pituitary cell column. Lower panel: concentration of rat prolactin in eluate from column. [Reproduced from ref. 54 with permission.]

influx into the cell and thus regulates release of prolactin from the cell. Our studies have been designed to test the effects of several agents on prolactin release in the light of these observations and hypothesis.

Our technique in the pituitary cell column allows us to follow dynamics of the release process of prolactin at minute-to-minute intervals. Like the observations of changes in electrical activity by dopamine at anterior pituitary cells, it might be expected that the effects of dopamine on prolactin release may be rapid. In figure 6 an inhibitory effect of dopamine was observed 3 min after the commencement of perfusion of the cells with dopamine, but 3 min before peak levels of dopamine were achieved. We purposely chose a supramaximal dose of dopamine (5 µM) to be certain that we would not be confused by an inadequate dose. At the time that peak dopamine levels were achieved prolactin levels were already suppressed by 45% of baseline and only a further 20% suppression was observed 5 min later. 3 min following the withdrawal of dopamine, prolactin was still maximally inhibited, but 1 min later, prolactin levels had doubled and 2.5 min later still prolactin levels had recovered to 80% of those observed before exposure to dopamine. From these results it is clear that the effects of dopamine are rapid and are rapidly reversible. However, the effects of dopamine on action potentials occur within seconds and thus the effects on release are slower.

The problem of understanding the mechanism by which cells respond to extracellular factors is central to the understanding of the control of secretion. The concept of the second messenger versus the coupling factors has been discussed by *Rasmussen and Goodman* [46]. They define a second messenger as 'a chemical signal, generated by interaction of an external stimulus with the cell surface receptor, which passes into one or more subcellular compartments and therein initiates the appropriate cellular response'. The lactotroph is unique in that it functions maximally in the absence of extracellular stimulatory factors; however, it is under tonic hypothalamic inhibitory control. Thus, the concept of a first, let alone second messenger, or coupling factor is problematic for the lactotroph. The concept of intrinsic calcium spikes acting as the driving force for the tonic secretion by the lactotroph is extremely attractive. These spikes may control the influx of calcium ions into the cell to permit prolactin secretion.

In spite of the wide knowledge of the specificity of certain drugs for stimulating and blocking the dopamine receptor on the lactotroph, the events that take place following stimulation of these receptors are unknown. Stimulation of these receptors has been described to stimulate adenylate cyclase activity [1] while other reports have suggested a reduction of adenylate cyclase activity [37]. However, the reduction of calcium spiking by dopamine observed by *Taraskevich and Douglas* [51] suggest that calcium influx may be intimately involved in the control of prolactin release and may indeed be the second messenger; thus, dopamine may inhibit spiking which may prevent influx of calcium into the cell and thereby inhibit prolactin secretion.

In order to establish the concept of calcium acting as the second messenger the concentrations of free calcium ions in the cytosol of the lactotrophs should be directly measured; its concentration should parallel hormone release. At the present time, techniques are not available to directly measure free calcium in the cytosol. Further, drugs which prevent calcium influx into

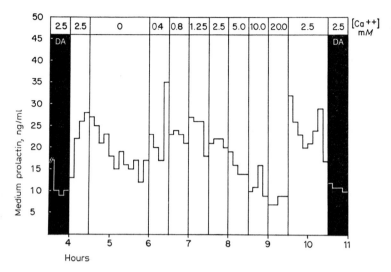

Fig. 7. The effects of varying extracellular calcium concentrations on prolactin release from pituitary cell column. Following 0.5-hour dose of 5 μM dopamine, the cells were exposed to calcium-free medium for 1.5 h and then 0.5-hour doses of increasing extracellular calcium concentrations. Following the restoration of normal extracellular calcium (2.5 mM at 9½ h) secretion was studied in the presence and absence of 5 μM dopamine. [Reproduced from ref. 54 with permission.]

the cell should inhibit prolactin release, and drugs which release calcium from intracellular stores should be capable of stimulating prolactin release, even in the absence of extracellular calcium.

If calcium is the 'messenger' or coupling agent, then lowering the extracellular calcium concentration should reduce influx of calcium into the cell and thereby reduce the calcium concentration in the cytosol. In figure 7 the effects of varying the extracellular calcium concentration are shown. The removal of extracellular calcium led to progressive reduction in prolactin release over 30 min. When the cells were exposed to increasing doses of calcium the release of prolactin increased; but when the dose of calcium was increased above physiological levels (2.5 mM) a dose-related inhibition of prolactin release was observed, which could be reversed by restoring the extracellular calcium concentrations to physiological levels. Similar observations of paradoxical suppression of release of secretory product have been made by *Douglas and Poisner* [17] at the posterior pituitary. At these sites stimulation of release was essential to demonstrate these effects. The observation of inhibition of spontaneous release of prolactin by high calcium underlines the intrinsic tonic nature of prolactin release, and again supports the hypothesis of spontaneous depolarizations of the lactotroph. The paradoxical effects of high calcium-inhibiting prolactin secretion are in contrast to our previous report of stimulation of prolactin release by 7.5 mM calcium. However, those studies did not examine the effects of higher calcium concentrations and used the hemipituitary incubation system [33].

If influx of extracellular calcium is responsible for the release process, the calcium may enter the cell by one of several channels in the cell membrane; through sodium channels, or through one or both of two types of calcium channels – the potential dependent or independent ones [46]. We first investigated the effect of two calcium channel-blocking agents manganese and a methoxy derivative of verapamil, D-600. Manganese is relatively specific for calcium channels alone, while D-600 probably blocks both sodium and calcium channels [46]. Sodium channels, however, appear unimportant in the control of prolactin release. In an unreported study, we confirmed *Parsons'* [43] observation that withdrawal of extracellular sodium had no effect on prolactin release. We have found tetrodotoxin, a selective sodium channel-blocking agent, to be ineffective in altering either spontaneous prolactin secretion or the inhibitory effects of dopamine on the release process [54]. These results are consistent with the lack of effect of removing extracellular sodium or tetrodotoxin on action potentials of the anterior pituitary.

In further experiments where cells were exposed over 4 h to calcium-free medium the prolactin release was inhibited even further (fig. 8). Following 2.5 h of exposure of the cells to calcium-free medium a supramaximal dose of dopamine (5 μM) had a negligible further effect on prolactin release, suggesting that the release process was already almost inhibited completely. On reexposure of the cells to physiological concentrations of extracellular calcium there was massive release of prolactin and the cells once more were sensitive to dopamine. Both manganese (2 mM) and D-600 (1 μM) were both effective in inhibiting spontaneous prolactin release. However, following withdrawal of manganese prolactin secretion recovered immediately while the effect of D-600 was prolonged. These results are consistent with those of *Takahara et al.* [49] showing that the calcium channel-blocking agent nifedipine also inhibited spontaneous prolactin release *in vitro*.

The hypothesis that tonic prolactin secretion is maintained by influx of extracellular calcium and that dopamine inhibits this process is supported by some experiments which we performed using the calcium ionophores which enhance calcium transport across membranes. The ionophore A23187 had no effect at doses from 1 nM to 10 μM on tonic prolactin secretion in the

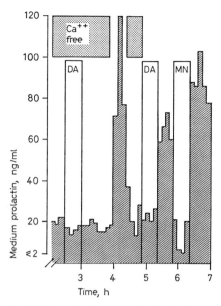

Fig. 8. Rat prolactin concentration in the eluate from pituitary cell column exposed to calcium-free medium in the presence or absence of dopamine (5 μM) and then medium containing normal calcium concentration and either dopamine (5 μM), medium alone, medium containing manganese (2 mM) followed by medium alone each for 0.5 h, 7.5-min fractions were collected. [Reproduced from ref. 54 with permission.]

presence of 2.5 mM calcium. However, it overcame, in a dose-dependent fashion, the inhibitory effects of dopamine [54].

The possibility that intracellular calcium may be physiologically important is suggested by the rather slower and less profound inhibitory effect of removing extracellular calcium, when compared to the effects of dopamine on prolactin release. This observation may arise from incomplete removal of extracellular calcium, since no chelating agent was used in these experiments. However, although traditionally the role of biogenic amines has been considered to be on the cell surface there is evidence that they may be taken up into the cell and may also have intracellular effects [4, 40]. Thus, dopamine may not only prevent influx of calcium from the extracellular fluid, but may also prevent mobilization of calcium from intracellular pools; thus, dopamine may inhibit prolactin release by lowering the cytosol calcium by effects at two different sites. We have evidence that intracellular calcium may be sufficient, under pharmacological conditions, to cause prolactin release. In the absence of extracellular calcium both ionophores X537A and A23187 were able to release prolactin [54]. It is possible that trace quantities of extracellular calcium may have been sufficient to sustain this release in the presence of the ionophores. However, in several other systems, including the β-cell of the pancreas and the gonadotrope of the pituitary, these ionophores even in the presence of chelating agents have been able to stimulate hormone release [5, 11]. In the parathyroid they were both effective in mobilizing intracellular calcium stores [6]. In the salivary gland, *Lowenstein* [29] demonstrated the release of intracellular calcium by both ionophores in the absence of extracellular calcium by use of intracellular injection of aequorin, a protein which fluoresces when combined with calcium ions.

cAMP and Prolactin Release

The role of cAMP in the control of prolactin release is confused. We have performed studies to reinvestigate this problem.

Effects of Dopamine on cAMP in Eluate from Pituitary Cell Column. Using the pituitary cell column technique we measured both prolactin and the cAMP concentration in the eluate from the column before, during and after withdrawal of 5 μM dopamine. Although the expected reversible inhibitory effects of dopamine on prolactin release were observed, there were no changes in the cAMP levels. As will be seen below we have evidence to suggest the cAMP in the eluate reflects changes of intracellular cAMP.

Effects of Cholera Toxin and Prostaglandin E_1 on Prolactin Release. Cholera toxin is effective in stimulating adenylate cyclase in all eukaryotic mammalian cells [21]. In spite of stimulating adenylate cyclase, documented by a rise in cyclic AMP in the eluate from the pituitary cell column, and observing the expected stimulation of growth hormone secretion neither tonic prolactin secretion nor the inhibiting effects of DA on prolactin release were affected (fig. 9). Similar results were observed with prostaglandin E_1.

Effects of Theophylline and Dopamine on Prolactin Release and Pituitary Cyclic AMP. The effects of theophylline were studied in the presence and absence of dopamine (table I). 10 mM theophylline had no effect on basal prolactin release, but completely reversed the inhibiting effects of 250 nM dopamine.

The cAMP content of the control and dopamine exposed glands were not different. Theophylline exposed glands demonstrated the expected rise

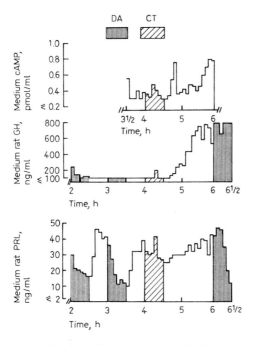

Fig. 9. The effect of 0.5-hour doses of 5 μM dopamine (DA) and 1 nM cholera toxin (CT) on the concentration of cyclic AMP, growth hormone (GH) and prolactin (PRL) in the eluate from pituitary cell column. [Reproduced from ref. 54 with permission.]

Table I. Effect of theophylline on the dopamine-mediated inhibition of prolactin release and anterior pituitary cAMP content

	μg medium prolactin/mg pituitary	anterior pituitary cAMP content pmol/mg protein
1 Control	7.8 ± 0.25	2.75 ± 0.50
2 Dopamine (250 nM)	5.1 ± 0.55[a]	2.35 ± 0.35
3 Theophylline (10 mM)	8.1 ± 0.90	4.99 ± 0.21[a]
4 Dopamine and theophylline	7.9 ± 0.50	2.49 ± 0.21[b]

[a] $p < 0.05$ vs control.
[b] n.s. vs control.

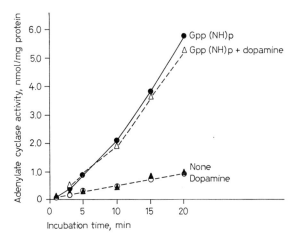

Fig. 10. Adenylate cyclase activity in pituitary homogenates after *in vitro* incubation with 100 μM Gpp(NH)p and/or 50 μM dopamine. [Reproduced from ref. 54 with permission.]

in cAMP, but the glands exposed to theophylline and dopamine together, had cyclic AMP content identical to the control – i.e. dopamine prevented the accumulation of cAMP expected after theophylline.

Effects of Dopamine on Basal and Gpp(NH)p Stimulated Anterior Pituitary Adenylate Cyclase Activity. Since dopamine had no effect on the cAMP levels in the eluate from the pituitary cell column or in the homogenate of anterior pituitary glands following *in vitro* incubations, the direct effects of dopamine on basal and Gpp(NH)p stimulated adenylate cyclase activity in pituitary homogenates were studied. Gpp(NH)p is a non-hydrolyzable analog of GTP and activates adenylate cyclase. Neither basal nor Gpp(NH)p (100 μM)-stimulated adenylate cyclase activity were effected by *in vitro* addition of 50 μM dopamine to the homogenates (fig. 10).

Effects of Dibutyryl cAMP. Dibutyryl cAMP as well as the nucleosides adenosine and guanosine have been previously shown by this laboratory to stimulate prolactin release [23].

Our results on the role of cAMP on the control of prolactin release initially appear to be conflicting. However, we believe cAMP to be relatively unimportant in the control of prolactin release. Dopamine is effective in inhibiting prolactin release, irrespective of adenylate cyclase activity. Dopamine does not alter the cAMP content of anterior pituitary cells and does not effect basal or Gpp(NH)p-stimulated adenylate cyclase activity. The only evidence in favor of a possible role of cAMP in prolactin release comes from the studies using theophylline and dbcAMP. Although we do not have a full explanation, their effects may be explained by other known effects of these two compounds. Theophylline is recognized in other cell types to stimulate release of intracellular calcium [5, 45]. Thus, both its lack of effect on tonic prolactin secretion (in spite of increasing cyclic AMP content of the pituitary), and its ability to reverse dopamine's inhibitory effect on prolactin release is similar to the effects observed with the ionophore. Why should dopamine inhibit the accumulation of cAMP in response to theophylline? If dopamine inhibits influx of extracellular calcium it will reduce cytosol calcium and thus might inhibit adenylate cyclase.

Similarly the effects of dbcAMP may be independent of its cAMP mimetic effects. In studies on parotid gland secretion, *Butcher* [7] has presented evidence that some component of the stimulation of amylase secretion by dbcAMP is mediated by an effect of dbcAMP on calcium, rather than by a cAMP mimetic effect. This may also be true at the beta cell of the pancreatic islet [5].

Although much work over the past 30 years has emphasized the role of intracellular calcium, more recently it has become clear that calcium may not act directly to regulate intracellular processes, but may act after binding to a protein or series of proteins which appear to be ubiquitous in eukaryotic cells – calmodulin or calcium-dependent regulators [10]. Calmodulin has four binding sites for calcium. Only after calcium has been bound to this protein can calcium, at least for certain processes, perform its regulatory role. Thus, the effects discussed in this paper may be only indirectly mediated by calcium alone but instead by calcium bound to calmodulin. The exciting concept of calmodulin acting as the calcium receptor is particularly relevant to the control of prolactin release since evidence may already be available to support it; we, as well as other investigators, have described that several neuroleptic drugs which, at low concentrations act as dopamine blocking drugs, at higher concentrations act as both inhibitors of prolactin release [16, 28, 57]. Initially it was suggested that they may act as partial dopamine agonists at these concentrations to inhibit prolactin release. However, this does not now appear likely. Furthermore, at these concentrations they are effective in inhibiting LH release induced by the gonadotropin-releasing hormone which probably is also normally mediated by calcium [11, 16].

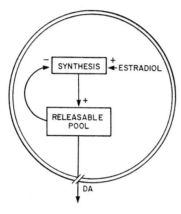

Fig. 11. Schematic representation of intracellular feedback mechanisms in the control of prolactin biosynthesis and release.

Conclusions

We, therefore, propose that normal tonic prolactin secretion is mediated by calcium-dependent action potentials which modulate intracellular calcium to maintain tonic prolactin secretion. This process may be inhibited physiologically by dopamine and experimentally by reduction in extracellular calcium, or by blockade of calcium channels by manganese or D-600 or finally by blockade of the receptors for calcium on calmodulin by neuroleptics at high dose. Prolactin secretion may also be stimulated by release of intracellular stores of calcium, for example, by dbcAMP. Adenylate cyclase and cAMP do not appear to be directly involved in the control of prolactin secretion.

The effects of dopamine, however mediated, are to rapidly control prolactin release. The degree of tonic inhibition is regulated by the concentration of dopamine to which the lactotrope is exposed. However, the secretory reserve appears to be regulated by two components: (a) the degree of dopaminergic inhibition; and (b) the effects of estrogen. The immediate effects of inhibition of release by dopamine or dopamine agonists is an accumulation of prolactin within the lactotroph, which in turn inhibits further prolactin biosynthesis. Estrogens appear to have their predominant effect on stimulating biosynthesis of prolactin, and this provides more prolactin which is available for release. This scheme is summarized in figure 11.

We have proposed this model which describes many diverse experimental observations, if for no other reason than that it is testable – hopefully in 1980 and beyond.

References

1. Ahn, A.S.; Gardner, E., and Makman, M.H.: Anterior pituitary adenylate cyclase: stimulation by dopamine and other monoamines. Eur. J. Pharmacol. *53:* 313–317 (1978).
2. Annunziato, L.: Regulatory mechanisms of tuberoinfundibular dopaminergic neurons; in MacLeod and Scapagnini, Central and peripheral regulation of prolactin function, pp. 62–71 (Raven Press, New York 1980).
3. Aubert, M.L.; Grumbach, S.L., and Kaplan, S.L.: The ontogenesis of human fetal hormones. J. clin. Invest. *56:* 155–164 (1975).
4. Berridge, M.J.: The role of 5-hydroxytryptomine and cyclic AMP in the control of fluid secretion by isolated salivary glands. J. exp. Biol. *53:* 171–186 (1976).
5. Brisson, G.R.; Malaisse-Lagae, F., and Malaisse, W.J.: The stimulus-secretion coupling of glucose-induced insulin release. J. clin. Invest. *51:* 232–241 (1972).
6. Brown, E.M.; Gardner, D.G., and Auerbach, G.D.: Effects of the calcium ionophore H23187 on dispersed bovine parathyroid cells. Endocrinology *106:* 133–138 (1980).
7. Butcher, F.: Calcium and cyclic nucleotides in the regulation of secretion from rat parotid by autonomic agonists; in George and Ignaro, Advances in cyclic nucleotide research, pp. 707–721 (Raven Press, New York 1978).
8. Calabro, M.A. and MacLeod, R.M.: Binding of dopamine to bovine anterior pituitary gland membrane. Neuroendocrinology *25:* 32–46 (1978).
9. Caron, M.C.; Beaulieu, M.; Raymond, V.; Gagne, B.; Drouin, J.; Lefkowitz, R.J., and Labrie, F.: Dopaminergic receptors in the anterior pituitary gland. J. biol. Chem. *253:* 2244–2253 (1978).
10. Cheung, W.Y.: Calmodulin plays a pivotal role in cellular regulation. Science *207:* 19–27 (1980).
11. Conn, P.M.; Rogers, D.C., and Sandhu, F.S.: Alteration of the intracellular calcium level stimulates gonadotropin release from cultured rat anterior pituitary cells. Endocrinology *105:* 1128–1134 (1979).
12. Copinschi, G.; L'Hermite, M.; Leclercq, R.; Golstein, J.; Vanhaelst, L.; Virasoro, E., and Robyn, C.: Effects of glucocorticoids on pituitary hormonal responses to hypoglycemia. Inhibition of prolactin release. J. clin. Endocr. Metab. *40:* 442–449 (1975).
13. Cramer, O.M.; Parker, C.M., Jr., and Porter, J.C.: Secretion of dopamine into hypophysial portal blood by rats bearing prolactin-secreting tumors or ectopic pituitary glands. Endocrinology *105:* 636–640 (1979).
14. Cronin, M.J.; Roberts, J.M., and Weiner, R.I.: Dopamine and dihydroergocryptine binding to the anterior pituitary and other brain areas of the rat and sheep. Endocrinology *103:* 302–309 (1978).
15. De Greef, W.J. and Neill, J.D.: Dopamine levels in hypophysial stalk plasma of the rat during surges of prolactin secretion induced by cervical stimulation. Endocrinology *105:* 1093–1099 (1979).
16. Denef, C; Van Nueten, J.M.; Leysen, J.E., and Janssen, P.A.J.: Evidence that pimozide is not a partial agonist of dopamine receptors. Life Sci. *25:* 217–226 (1979).
17. Douglas, W.W. and Poisner, A.M.: Stimulus-secretion coupling in a neurosecretory

organ: the role of calcium in the relase of vasopressin from the neurohypophysis. J. Physiol., Lond. *172:* 1–18 (1964).
18   Eikenburg, D.C.; Ravitz, A.J.; Gudelsky, G.A., and Moore, K.E.: Effects of estrogen on prolactin and tuberoinfundibular dopaminergic neurons. J. neural. Transm. *40:* 235–244 (1977).
19   Friesen, H.G.: Human prolactin. Ann. R. Coll. Phys. Surg. Canada *11:* 275–281 (1978).
20   Gautvick, K.M.; Tashjian, A.H., Jr.; Kourides, I.A.; Weintraub, B.D.; Graeber, C.T.; Maloof, F.; Suzuki, K., and Zuckerman, J.E.: Thyrotropin-releasing hormone, prolactin release and suckling. New Engl. J. Med. *290:* 1162–1165 (1974).
21   Gill, D.M.: Mechanism of action of cholera toxin. Adv. cyclic Nucl. Res. *8:* 85–118 (1977).
22   Heuson, J.C.; Waelbroeck, C.; Legros, N.; Gallez, G.; Robyn, C., and L'Hermite, M.: Inhibition of DMBA-induced mammary carcinogenesis in the rat by 2-Br-α-ergocryptine (CB 154), an inhibitor of prolactin secretion and by nafoxidine (U-11, 100A), an estrogen antagonist. Gynecol. Invest. *2:* 130–137 (1971/72).
23   Hill, M.K.; MacLeod, R.M., and Oraitt, P.: Dibutyryl cyclic AMP, adenosine and guanosine blockade of the dopamine, ergocryptine and apomorphine inhibition of prolactin release *in vitro*. Endocrinology *99:* 1612–1617 (1976).
24   Jacobi, J.M. and Lloyd, H.M.: Anti-oestrogenic effects of nafoxidine on the synthesis of DNA by the rat pituitary gland. J. Endocr. *76:* 555–556 (1978).
25   Jordan, V.C. and Koerner, S.: Tamoxifen as an anti-tumor agent: role of oestradiol and prolactin. J. Endocr. *68:* 305–311 (1976).
26   Krieger, D.T.; Howaritz, P.J., and Frantz, A.G.: Absence of nocturnal elevation of plasma prolactin concentrations in Cushing's disease. J. clin. Endocr. Metab. *42:* 260–272 (1976).
27   Labrie, F.; Ferland, L.; Dipaolo, T., and Veilleux, R.: Modulation of prolactin secretion by sex steroids and thyroid hormones; in MacLeod and Scapagnini, Central and peripheral regulation of prolactin function (Raven Press, in press).
28   Lamberts, S.W.J. and MacLeod, R.M.: The biphasic regulation of prolactin secretion by dopamine agonist-antagonists. Endocrinology *103:* 287–295 (1978).
29   Lowenstein, W.B.: Permeable junctions. Cold Spring Harb. Symp. quant. Biol. *40:* 49–63 (1976).
30   Lowry, P.J.: A sensitive method for the detection of corticotrophin releasing factor using a perfused pituitary cell column. J. Endocr. *62:* 163–164 (1974).
31   MacLeod, R.M.: Influence of norepinephrine and catecholamine-depleting agents in the synthesis and release of prolactin and growth hormone. Endocrinology *85:* 916–923 (1969).
32   MacLeod, R.M.: Regulation of prolactin secretion; in Martini and Ganong, Frontiers in neuroendocrinology, vol. 4, pp. 169–194 (Raven Press, New York 1976).
33   MacLeod, R.M. and Fontham, F.M.: Influence of ionic environment on the *in vitro* synthesis and release of pituitary hormones. Endocrinology *86:* 863–869 (1970).
34   MacLeod, R.M. and Lamberts, S.W.J.: Clinical and fundamental correlates in dopaminergic control of prolactin secretion; in Muller and Agnoli, Neuroendocrine correlates in neurology and psychiatry, pp. 89–101 (Elsevier/North-Holland Biomedical Press, Amsterdam 1979).

35 MacLeod, R.M. and Lehmeyer, J.E.: Regulation of the synthesis and release of prolactin; in Wolstenholme and Knight, Lactogenic hormones, pp. 53–82 (Churchill/Livingstone, London 1971).
36 MacLeod, R.M. and Lehmeyer, J.E.: Studies on the mechanism of the dopamine mediated inhibition of prolactin secretion. Endocrinology 94: 1077–1085 (1974).
37 Markstein, R.; Herrling, P., and Wagner, H.: Bromocriptine mimicks dopamine effects on the cyclic-AMP system of rat pituitary gland. Abstr. 7th Int. Congr. of Pharmacology, Paris (Pergamon Press, London 1978).
38 Nagy, I.; Valdenegro, C.A., and MacLeod, R.M.: Effect of anti-estrogens on pituitary prolactin production in normal and pituitary tumor-bearing rats. Neuroendocrinology (in press).
39 Nagy, I.; Valdenegro, C.A.; Login, I.S.; Snyder, F.A., and MacLeod, R.M.: Estradiol and anti-estrogens in pituitary hormone production, metabolism, and pituitary tumor growth; in Jacobelli, 1st Int. Congr. on Hormones and Cancer, Rome (Raven Press, in press, 1979).
40 Nansel, D.D.; Gudelsky, G.A., and Porter, J.C.: Subcellular localization of dopamine in the anterior pituitary gland of the rat: apparent association of dopamine with prolactin secretory granules. Endocrinology 105: 1073–1077 (1979).
41 Nicholson, R.I. and Golder, M.P.: The effect of synthetic anti-oestrogens on the growth and biochemistry of rat mammary tumours. Eur. J. Cancer 11: 571–579 (1975).
42 Noel, G.L.; Suh, H.K., and Frantz, A.G.: Prolactin release during nursing and breast stimulation in post partum and non post-partum subjects. J. clin. Endocr. Metab. 38: 413–423 (1974).
43 Parsons, J.A.: Effects of cations on prolactin and growth hormone secretion by rat adenohypophyses in vitro. J. Physiol., Lond. 210: 973–987 (1970).
44 Perkins, N.A.; Westfall, T.C.; Paul, C.V.; MacLeod, R.M., and Rogol, A.D.: Effect of prolactin on dopamine synthesis in medial basal hypothalamus: evidence for short loop feedback. Brain Res. 160: 431–444 (1979).
45 Poisner, A.M.: Direct stimulant effect of aminophylline on catecholamine release from the adrenal medulla. Biochem. Pharmacol. 22: 469–476 (1973).
46 Rasmussen, H. and Goodman, D.B.P.: Relationships between calcium and cyclic nucleotides in cell activation. Physiol. Rev. 57: 421–509 (1977).
47 Seo, H.; Refetoff, S.; Martino, E.; Vassart, G., and Brocas, H.: The differential stimulatory effect of thyroid hormone on growth hormone synthesis and estrogen on prolactin synthesis due to accumulation of specific messenger ribonucleic acids. Endocrinology 104: 1083–1090 (1979).
48 Shupnik, M.A.; Baxter, L.A.; French, L.R., and Gorski, J.: In vivo effects on estrogen on ovine pituitaries: prolactin and growth hormone biosynthesis and messenger ribonucleic acid translation. Endocrinology 104: 729–735 (1979).
49 Takahara, J.; Yunoki, S.; Yamane, Y.; Yamauchi, J.; Yakushiji, W.; Kageyama, N.; Fujino, K., and Ofuji, T.: Effects of nifedipine on prolactin, growth hormone, luteinizing hormone release by rat anterior pituitary in vitro. Proc. Soc. exp. Biol. Med. 162: 31–33 (1979).
50 Taraskevich, P.S. and Douglas, W.W.: Action potentials occur in cells of the normal anterior pituitary gland and are stimulated by the hypophysiotropic peptide thyrotropin-releasing hormone. Proc. natn. Acad. Sci. USA 74: 4064–4067 (1977).

51 Taraskevich, P.S. and Douglas, W.W.: Catecholamines of supposed inhibitory hypophysiotrophic function suppress action potentials in prolactin cells. Nature, Lond. *276:* 832–834 (1978).
52 Thorner, M.O.: Prolactin: clinical physiology and the significance and management of hyperprolactinemia; in Martini and Besser, Clinical neuroendocrinology, pp. 319–361 (Academic Press, New York 1977).
53 Thorner, M.O.; Round, J.; Jones, A.; Fahmy, D.; Groom, G.V.; Butcher, S., and Thompson, K.: Serum prolactin and oestradiol levels at different stages of puberty. Clin. Endocrinol. *7:* 463–468 (1977).
54 Thorner, M.O.; Hackett, J.T.; Murad, F., and MacLeod, R.M.: Calcium rather than cAMP as the physiological intracellular regulator of prolactin release (Neuroendocrinology, in press).
55 Turkington, R.W.; Underwood, L.E., and Van Wyk, J.J.: Elevated serum prolactin levels after pituitary stalk section in man. New Engl. J. Med. *285:* 707–710 (1971).
56 Vician, L.; Shupnik, M.A., and Gorski, J.: Effects of estrogen on primary ovine pituitary cell cultures: stimulation of prolactin secretion, synthesis, and preprolactin messenger ribonucleic acid activity. Endocrinology *104:* 736–743 (1979).
57 West, B. and Dannies, P.S.: Antipsychotic drugs inhibit prolactin release from rat anterior pituitary cells in culture by a mechanism not involving the dopamine receptor. Endocrinology *104:* 877–880 (1979).
58 Yeo, T.; Thorner, M.O.; Jones, A.; Lowry, P.J., and Besser, G.M.: The effects of dopamine, bromocriptine, lergotrile and metoclopramide on prolactin release from continuously perfused columns of isolated rat pituitary cells. Clin. Endocrinol. *10:* 123–130 (1979).

M.O. Thorner, MB, Department of Internal Medicine, Division of Endocrinology and Metabolism, University of Virginia School of Medicine, Charlottesville, VA 22908 (USA)

Discussion

Herschman: Could you amplify the evidence for dibutyryl cyclic AMP acting through the mechanism of calcium secretion or calcium influx?

Thorner: Butcher, on the salivary gland, using various analogues of cyclic AMP and cyclic GMP, showed that only dibutiryl cyclic AMP's effect was dependent on whether the cells had been exposed to low calcium concentration. So, part of the effect of dibutyryl cyclic AMP was abolished if the cells had been exposed to calcium-free medium. With the other analogues, there was an effect which was only a proportion of that seen with dibutyryl cyclic AMP, and that was completely unaffected, irrespective of whether the cells had been exposed to normal calcium or not.

Eastman: Could you amplify the effect you showed of haloperidol on calmodulin?

Thorner: It was done *in vitro,* and explants of the rat pituitary were exposed *in vitro* to these doses. The dose used normally as a blocking dose of haloperidol is about 10^{-7} mol and the dose here was 10^{-3} mol, so it is four orders of magnitude greater. I am sure it is a pharmacological effect.

Wuttke: In one of your slides, synthesis appeared to be stimulated by bromocriptine.

Thorner: It is new synthesis that is being studied. There is little doubt that bromocriptine can inhibit the new synthesis of prolactin. Some of the discrepancies that occur may well be due to the duration for which the animals are exposed to bromocriptine. In the first experiment, we were seeing the glands removed from animals which had been treated *in vivo* with bromocriptine. In the second experiment the bromocriptine was added *in vitro*. You are also, therefore, looking at a time effect.

Jacobs: Do you not agree that time is a critical determinant, even *in vitro*? If one adds dopamine or dopamine agonist *in vitro*, and one looks only at new synthesis and new release over the very short-term, one can see very clearly inhibition of secretion with no change in total synthesis, when one sums the quantity in the gland and in the medium. If one continues such incubations beyond, say, 3–4 h, not only is release inhibited, but synthesis is also suppressed. It is very hard, for me at least, to avoid the suspicion that the same events might occur no matter what agent one used as an effective inhibitor of secretion. That is, does the inhibition of secretion by itself lead, ultimately, to the suppression of synthesis, so that the lactotrope cell has some kind of an autoregulatory system? When secretion is potently inhibited, stores do build up in the gland, and at early time-points this can be clearly demonstrated.

Thorner: I have the same feeling as you, that it is the release process which is primarily being inhibited, which leads to autofeedback within the cell. Oestrogen is exactly the reverse, it stimulates the synthesis, and then there is autofeedback again.

Characteristics of Prolactin Secretion from Anterior Pituitary Implants[1]

Robert A. Adler[2], Susan J. Brown and Hilda Weyl Sokol

Departments of Medicine and Physiology, Dartmouth Medical School, Hanover, N.H.

For many years hyperprolactinemia has been induced by implantation of an animal's hypophysis or donor pituitary glands under the kidney capsule [4]. Removal of the anterior pituitary from prolactin (PRL) inhibitory factor(s) has been demonstrated to cause increased secretion of PRL from the ectopic glands. *Sinha and Tucker* [8] demonstrated mammary stimulation by ectopic pituitaries, and this model has been used to study the effects of chronic hyperprolactinemia. Is it a valid one?

Material and Methods

Four to eight anterior pituitary glands from littermates were placed under the kidney capsules of weanling Sprague-Dawley male rats. Sham-operated rats received muscle tissue. For blood drawing, many rats had a polyethylene catheter placed through the carotid artery into the descending aorta. This catheter was kept patent with heparinized saline. Some rats were exsanguinated by aortic puncture after ketamine and pentobarbital anesthesia. At sacrifice the kidneys and one half of the intrasellar pituitary were fixed in Bouin's solution. The other half pituitary was immediately frozen. Sera and pituitary homogenates were assayed for PRL concentration using NIAMDD reagents with a slight modification of the NIAMDD protocol. After imbedding in paraffin, implants and intrasellar pituitaries were studied using the immunoenzyme method of *Sternberger* [9]. The following antisera were used (kindly supplied by NIAMDD): anti-rat PRL, anti-rat growth hormone (rGH), anti-rat thyroid-stimulating hormone (rTSH), anti-rat luteinizing hormone (rLH), and anti-rat follicle-stimulating hormone (rFSH).

1-ml samples of serum from 5 hyperprolactinemic rats were placed on 1.5×90 cm columns of Sephadex G-100. 2-ml fractions were collected and assayed for PRL immunoreactivity. All data were analyzed using Student's t-test.

[1] Supported by NIH Grant AM22032.
[2] Dr. Adler is the recipient of NIH Clinical Investigator Award AM00707.

Results

Serum PRL levels in implanted animals ranged from 80 to over 300 ng/ml; most animals with 4 implants had PRL concentrations of approximately 100 ng/ml and those with 8 implants had PRL concentrations of 200 ng/ml or greater. Depending on method of blood drawing, and therefore amount of stress (or presence of anesthesia), sham rats had PRL levels of 5–50 ng/ml. Unanesthetized sham rats with aortic catheters had the lowest PRL values. The mean serum PRL concentrations at sacrifice for the groups of animals studied for intrasellar pituitary PRL concentration are shown in table I.

The anterior pituitary glands implanted under the kidney capsule demonstrated a marked reaction with anti-rPRL using the immunoenzyme technique, as shown in figure 1a. The cells had much granular cytoplasm. There was some 'staining' with anti-rGH, but the cells were small and had only a rim of cytoplasm (fig. 1b). Adjacent sections treated with anti-rTSH (fig. 1c), anti-rLH (fig. 1d), and anti-rFSH (fig. 1e) showed little reaction.

When the immunoenzyme technique was applied to sections of the intrasellar pituitaries, reaction with anti-rPRL was considerably greater in the pituitaries from sham-operated rats. Figure 2a shows the intrasellar pituitary reaction with anti-rPRL in implanted rat 3-4-3. In contrast the intrasellar pituitary from sham rat 3-2-4 demonstrates considerably more reaction product to anti-rPRL, as shown in figure 2b. In table II the intrasellar pituitary PRL concentration in these two rats parallels the immunoenzyme reaction. The intrasellar pituitary PRL concentration was significantly lower in the chronically hyperprolactinemic rats as a group (table I).

Table I. Serum and pituitary prolactin concentrations in implanted and sham-operated rats

	Serum prolactin ng/ml ± SEM	Pituitary prolactin ng/mg wet weight ± SEM
Implanted (n = 14)	180 ± 18[1]	420 ± 39[2]
Sham-operated (n = 11)	44 ± 5	675 ± 78

[1] Different from shams at $p < 0.001$.
[2] Different from shams at $p < 0.015$.

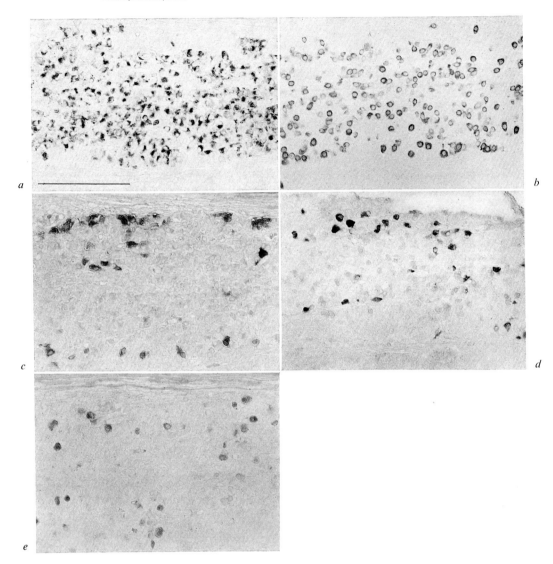

Fig. 1. Implanted pituitaries were examined using the immunoenzyme technique of *Sternberger* [9]. *a* Reaction with anti-rPRL; *b* reaction with anti-rGH; *c* reaction with anti-rTSH; *d* reaction with anti-rLH; *e* reaction with anti-rFSH. Scale = 100 μm.

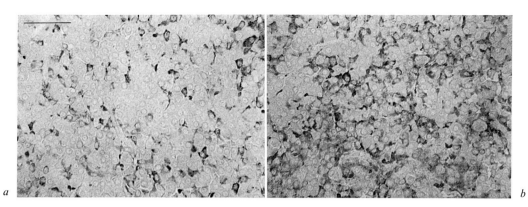

a   b

Fig. 2. The intrasellar pituitary glands were examined using the immunoenzyme technique with anti-rPRL. *a* Intrasellar pituitary of implanted rat 3–4–3; *b* intrasellar pituitary of sham-operated rat 3–2–4. Scale = 50 μm.

Table II. Serum and pituitary prolactin concentrations in representative rats (fig. 2a, b)

Rat number	Imp./sham	Serum PRL (ng/ml): no anesthesia	Serum PRL (ng/ml): ketamine + pentobarbital	Intrasellar pituitary PRL ng/mg wet weight
3–4–3	implanted	230	254	300
3–2–4	sham-operated	31	46	788

Table III. Immunoreactivity of hyperprolactinemic rat serum after chromatography with Sephadex G-100 (means ± SEM)

	Serum PRL ng/ml	Recovery %	Peak I %	Peak II %	Peak III %
Implanted rats (n = 5)	164.3 ± 24.5	95.7 ± 4.8	11.8 ± 1.2	8.2 ± 1.5	75.9 ± 1.4

When serum samples from hyperprolactinemic rats were chromatographed on Sephadex G-100, approximately 76% of the immunoreactivity was found in peak III, which correponds to NIH rPRL. Table III lists the means of 5 similarly chromatographed rat sera. Recovery ranged from 85 to 105%.

Discussion

This study confirms previous work showing feedback on the hypothalamo-pituitary unit by other sources of PRL. *Chen et al.* [3] have shown that PRL-secreting tumors could decrease intrasellar pituitary PRL contents estimated by pigeon crop assay. Similar findings, also using bioassay techniques for PRL, have been noted for cocoa butter implants [10] and for implanted pituitaries [7]. We have shown previously by RIA [1] that implanting extra pituitaries in an unusual strain of rat (the Brattleboro strain of Long-Evans rat with hereditary hypothalamic diabetes insipidus) decreased intrasellar pituitary PRL content. We have now extended this by demonstrating similar feedback in an endocrinologically normal rat strain and have correlated pituitary PRL measurements with immunoenzyme 'staining'. Cells reacting with anti-rPRL were considerably more numerous in the intrasellar pituitaries of our sham-operated rats.

Conversely the implants themselves showed marked reaction with anti-rPRL antiserum. The cells were large and had much granular cytoplasm. In contrast, although there were fairly numerous somatotrophs present in the implants, the cells that reacted with anti-rGH were small and had little cytoplasm. They clearly did not resemble intrasellar somatotrophs in overall architecture. Implant cells reacting with anti-rTSH, anti-rLH, and anti-rFSH were generally small in number and size. We did not assess ACTH and related peptides in the implanted rats. However, *Kendall et al.* [6] showed many years ago that hypophysectomized rats with several implants had corticosterone levels much lower than normal rats. Furthermore, *Bartke et al.* [2] showed no difference in serum corticosterone between non-hypophysectomized implanted and sham rats. Nonetheless, the adrenal glands were larger in *Bartke*'s implanted rats, perhaps reflecting increased stimulation of androgen production. Further study of ACTH and adrenal steroid production is required in the anterior pituitary implanted rat model.

Our data also indicate that most of the PRL secreted by the implants is of the molecular size corresponding to native rPRL (peak III). This has recently been shown to have greatest radioreceptor activity [5]. About 20% of the remaining immunoreactivity eluted at two different fractions corresponding to so-called 'big prolactin' and 'big, big' or void-volume PRL. This finding suggests that not only do the hypothalamus and intrasellar pituitary of the implanted rats react to an active PRL, but the circulating PRL is also of the proper molecular size for activity.

The data reported therefore suggest that the rat with extra anterior pituitary glands under the kidney capsule is a valid model of chronic hyperprolactinemia.

References

1 Adler, R.A.; Dolphin, S.; Szefler, M., and Sokol, H.W.: The effects of elevated circulating prolactin in rats with hereditary hypothalamic diabetes insipidus (Brattleboro strain). Endocrinology *105:* 1001–1006 (1979).
2 Bartke, M.S.; Smith, M.S.; Michael, S.D.; Peron, F.G., and Dalterio, S.: Effects of experimentally-induced chronic hyperprolactinemia on testosterone and gonadotropin levels in male rats and mice. Endocrinology *100:* 182–186 (1977).
3 Chen, C.L.; Minaguchi, H., and Meites, J.: Effects of transplanted pituitary tumors on host pituitary prolactin secretion. Proc. Soc. exp. Biol. Med. *126:* 317–320 (1967).
4 Everett, J.W.: Functional corpora lutea maintained for months by autografts of rat hypophyses. Endocrinology *58:* 786–796 (1956).
5 Farkouh, N.H.; Packer, M.G., and Frantz, A.G.: Large molecular size prolactin with reduced receptor activity in human serum: high proportion in basal state and reduction after thyrotropin-releasing hormone. J. clin. endocr. Metab. *48:* 1026–1032 (1979).
6 Kendall, J.W.; Stott, A.K.; Allen, C., and Greer, M.A.: Evidence for ACTH secretion and ACTH suppressibility in hypophysectomized rats with multiple heterotopic pituitaries. Endocrinology *78:* 533–537 (1966).
7 Mena, F.; Maiweg, H., and Grosvenor, C.E.: Effect of ectopic pituitary glands upon prolactin concentration by the *in situ* pituitary of the lactating rat. Endocrinology *83:* 1359–1362 (1968).
8 Sinha, Y.N. and Tucker, H.A.: Pituitary prolactin and mammary development after chronic administration of prolactin. Proc. Soc. exp. Biol. Med. *128:* 84–88 (1968).
9 Sternberger, L.A.: Immunocytochemistry, pp. 1–184 (Prentice-Hall, Englewood Cliffs 1974).
10 Voogt, J.L. and Meites, J.: Effects of an implant of prolactin in median eminence of pseudopregnant rats on serum and pituitary LH, FSH and prolactin. Endocrinology *88:* 286–292 (1971).

R.A. Adler, MD, Endocrine/Metabolism Division, Department of Medicine, Dartmouth Medical School, Hanover, New Hampshire 03755 (USA)

Discussion

Bern: You might have a chance here also to see whether increasing prolactin levels have any effect on your pituitary transplants. Is there any direct effect on the prolactin cell itself which is not mediated through the hypothalamus? Have you considered that as a possible study subject?

Adler: We have not done that.

Neill: A group from San Antonio claimed that there were direct inhibitory effects of prolactin on the pituitary for the secretion of prolactin.

Leong: The effects of prolactin on the pituitary itself have been quite well established and there is increasing *in vitro* evidence that prolactin does have an effect on the pituitary, especially at a very high level.

Herschman: In contrast to what we have just heard, I think that the evidence of prolactin directly inhibiting secretion of the lactotrope is unclear. We have data showing that the GH3 cell incubated with ovine prolactin at relatively high concentration continued, in short-term incubation, to secrete rat prolactin in an undisturbed manner. Thus, the ovine prolactin did not directly feed back, inhibiting secretion of rat prolactin in this *in vitro* system.

Wuttke: I would totally agree with that last statement. We also co-incubated ovine prolactin with pituitary halves, and did not see any effect on the prolactin secretion. I would agree that it is not clear at all whether prolactin has a direct action at the pituitary.

Greenwood: Has anyone detected prolactin receptors on the pituitary? Or is the implication that the prolactin is going in directly by some unknown mechanism?

Kelly: About 3 years ago, *William Frantz* from Michigan State had a report in *Science* on measuring prolactin receptors in pituitary cells in culture – rather fantastically high levels of prolactin receptors as a matter of fact. We tried to repeat that study in our laboratory, and failed to find any significant binding in pituitary cells in culture. We have since used a technique to remove endogenous prolactin, because there is a large amount of prolactin, of course, which would be occupying the receptors, if they are there. Using magnesium chloride to remove endogenous prolactin bound to the receptors, we are able to measure very small levels of prolactin receptors: there is something there but it is very small in quantity.

Electrophysiological Study of Prolactin-Secreting Pituitary Cells in Culture

Effects of Thyrotropin-Releasing Hormone, Gonadal Steroids and Dopamine[1]

Bernard Dufy, Jean-Didier Vincent, Danièle Gourdji and Andrée Tixier-Vidal

Institut National de la Santé et de la Recherche Médicale U 176, Bordeaux, and Groupe de Neuroendocrinologie, Collège de France, Paris

The release of prolactin (PRL) by the pituitary gland depends largely upon the complex interaction of a number of nervous and hormonal influences. At the present time there is no definite identification of the different factors responsible for the control of PRL secretion. It is well established that the secretion of PRL is principally under a hypothalamic inhibitory influence [8, 20]. Anatomical and biochemical studies have clearly demonstrated that dopamine (DA) is a physiological inhibitor of PRL secretion [8, 20]; however, a number of other hypothalamic factors such as γ-aminobutyric acid (GABA) are also candidates for a PRL inhibiting activity [33, 34]; more recently, histidyl-proline-diketopiperazine, a degradation product of the thyrotropin-releasing hormone (TRH) has also been shown to have a PRL-inhibiting activity [3, 4]. On the other hand, various substances stimulate PRL secretion either directly at the pituitary level or indirectly through the hypothalamus. TRH stimulates directly the secretion of PRL by pituitary cells [27, 37, 38] but estrogens are also able to increase the production of PRL [7, 16, 43]. Despite the numerous works done in this field, the mechanisms of action of the PRL-inhibiting and stimulating factors are still poorly known. The general opinion is that calcium ions (Ca^{2+}) play a major role in the secretion process of pituitary hormones [26]. The stimulus-secretion coupling concept, proposed by *Douglas* [9] to explain the release of catecholamines and posterior pituitary hormones, may also apply to the release of several anterior pituitary hormones including PRL [13, 40].

---

[1] We thank *J.B. Seal* for his help with the manuscript. This work has been supported by grants from INSERM (U 176 and CRL 87-1-2656), DGRST (C 77.7.0654), the University of Bordeaux II and the CNRS (ERA 493 and ER 89).

Classically, a major difference between neurons and gland cells is the regenerative voltage-dependent conductance changes (action potentials) observed in the former and the passive membrane characteristics of the latter. In neurons, including neuroendocrine cells, action potentials provoke secretion by the opening of voltage-dependent calcium channels. However, the increase in intracellular Ca^{2+} concentration which accompanies secretion from gland cells has not been considered to be associated with action potentials. Such a division between excitable nerve cells and non-excitable gland cells is no longer tenable. To date some exceptions to this classical division have already been found: the β cells of the endocrine pancreas [23], the chromaffin cells of the adrenal medulla [5, 6] and pituitary cells from pars distalis [35] and pars intermedia [10].

The ionic requirements for action potentials in endocrine cells appear to be different to those for neurons and even between the various types of endocrine cells: β cells and pituitary cells of the pars distalis only require calcium whereas chromaffin cells and cells of the pars intermedia seem to require both Na^+ and Ca^{2+} to sustain spiking activity. However, spiking activity always results in activation of Ca^{2+} channels and hence an increase in intracellular Ca^{2+} concentration. This increase may, in turn, be a determinant factor in the mechanism of secretion.

The study of GH3/B6 cells, a subclone of the GH3 rat prolactin cell line [37], has shown that these cells display action potentials under certain conditions [11, 12]. Secretion of prolactin by this cell line is stimulated by TRH [38] and estradiol [41] but inhibited by dopamine [10]. These substances also modify the electrical activity of the cells [11, 12]. Hence the GH3/B6 cell line constitutes a good model for studying the relationship between electrical activity and secretion in prolactin cells.

Electrophysiological Characteristics of GH3/B6

In our studies GH3/B6 cells were routinely grown in Ham's F 10 solution supplemented with 15% horse serum and 2.5% fetal calf serum, as previously described [11]. Cells were cultured for 5–7 days before being used in experiments. For recordings, the culture medium was replaced by a bathing solution of the following composition (in mM per liter): NaCl 142.6; KCl 5.6; $CaCl_2$ 10; glucose 5 and Hepes buffer 5 (pH 7.4).

Stable intracellular recordings (20–30 min) were usually obtained and cells could sometimes be recorded for more than 1 h. The mean resting mem-

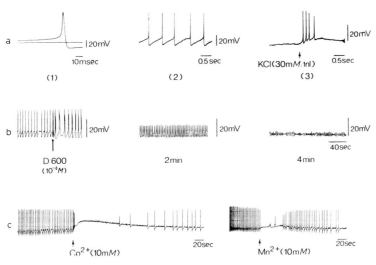

Fig. 1. Characteristics of action potentials recorded from GH3/B6 cells. *a* Spontaneous action potentials (1 and 2) and action potentials induced by the administration of KCl directly on the cell being recorded by a pressure injection system. *b* Effect of D 600, a blocker of Ca^{2+} channels, on shape and amplitude of action potentials in a spontaneously firing cell: (i) control; (ii and iii) respectively 2 and 4 min after administration of D 600 (10^{-4} M, 2nl). *c* Effects of cobalt (Co^{2+}) and manganese (Mn^{2+}) ions on spontaneously firing cells. It should be noted that Co^{2+} and Mn^{2+} completely suppressed action potentials and depolarized the cell which is consistent with an involvement of Ca^{2+} ions in the spiking activity and also the resting membrane polarization.

brane potential was 40 ± 5 mV (mean $\pm$ SD; n = 120). The mean input resistance was 185 MΩ. 50% of the cells appeared electrically excitable in that they displayed action potentials during a depolarizing intracellular current injection (0.3–0.7 nA, 0.5 sec).

Spikes were also obtained during depolarization induced by the application of KCl close to the cell, using a pneumatic ejection system (fig. 1). 28% of all cells were spontaneously active with a firing rate which never exceeded 2 Hz. The observation that only 50% of the cells were electrically excitable is puzzling. The resting potential and input resistance of the non-excitable cells (33 ± 3 mV; 115 ± 25 MΩ; n = 50) were significantly different from that of excitable cells, thus it is possible that the membrane of non-excitable cells was damaged. Experimental manipulations of the ionic composition of the bathing medium showed that the action potentials were calcium (Ca^{2+})-dependent; action potentials were still recorded in the presence of tetrodotoxin, a known blocker of sodium (Na^+) channels, and also

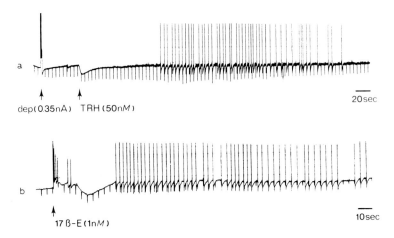

Fig. 2. Effects of TRH and 17β-estradiol (17β-E) on the membrane electrical activity of two GH3/B6 cells. Both substances were dissolved in the bathing solution. 17β-E was first dissolved in alcohol, the final alcohol concentration being 0.1%. *a* Ejection of TRH (50 n*M*, 2 nl) into the vicinity of the cell membrane elicits a precocious hyperpolarization followed within 1 min by a sustained spiking activity. *b* Ejection of 17β-E directly onto the membrane of the cell induces an early depolarization followed by the firing of action potentials. A measure of the membrane resistance (vertical bars) is obtained by injecting a hyperpolarizing current (0.16 nA). The solvents were ineffective.

in a Na^+ free medium. They were suppressed by cobalt (Co^{2+}), and D 600, a blocker of Ca^{2+} channels (fig. 1). These findings clearly indicate that opening of potential dependent calcium channels is responsible for a major part of the action potential. We therefore suggest that the function of action potentials in GH3/B6 may be to increase the intracellular free calcium. Since calcium has also been implicated in secretory activity, it is interesting to examine whether factors which influence secretion also have an effect on the calcium-dependent electrical activity of GH3/B6 cells.

Effects of TRH on the Electrical Activity of GH3/B6 Cells

TRH (25–125 n*M*) added to the bathing solution increased the percentage of cells displaying action potentials, which was consistent with the effect of TRH observed with extracellular recording [17, 35]. Ejection of TRH (50 n*M*, 2 nl) close to the cell evoked a train of action potentials within 1 min (fig. 2). This spiking activity was preceded by a progressive increase of the input resistance without any detectable change in the resting membrane polarization, apart from an early hyperpolarization which did not seem to be directly involved in the spiking activity. After 1–10 min the cell stopped firing as the resistance returned to the resting level. Once again, the action potentials were reduced or completely abolished by D 600. Furthermore, the action potentials were only slightly changed in Na⁺-free medium, although no overshoot was observed. This is consistent with the demonstration of the role of calcium in TRH-stimulated release of PRL by GH3 cells and is in agreement with the concept of a stimulus-secretion coupling mechanism.

The mechanism by which TRH induces spiking without changing the resting potential remains to be determined. TRH may possibly increase the excitability through lowering of spike threshold by changing the voltage at which activation of the voltage-dependent conductances underlying the spike occurs. Such a direct effect of a peptide in changing the action potential threshold and hence excitability has already been observed in spinal cord neurons [1]. Other examples of the action of peptides on voltage-dependent conductances have also been described for invertebrate neuron systems [2].

Early Membrane Permeability and Conductance Changes following Administration of Estradiol

We observed that 17β-estradiol (17β-E) when ejected close to a cell being recorded elicited a series of transient ionic permeability changes beginning within 1 sec after administration of 17β-E (fig. 2). The modifications of passive membrane properties were in two parts: a transient depolarization and a decrease of the overall input resistance which induced a short burst of action potentials. This was followed, 1 min later, by an increase in the input resistance which leads, in some excitable cells, to a sustained discharge of Ca^{2+}-dependent action potentials (fig. 2). When an ejection was successful in eliciting spikes a subsequent ejection of 17β-E 3–10 min later was totally ineffective. 17α-Estradiol had a much weaker effect even at higher doses. Progesterone and testosterone were not able to induce spiking activity.

These results reveal a rapide effect of 17β-E on the membrane of GH3/B6 cells. The electrical activity elicited by 17β-E was similar in time course and Ca^{2+} dependence to that induced by TRH. The rapid and specific effect of 17β-E on membrane properties implies recognition sites for the steroid at the membrane surface and probably reflects conformational changes in membrane components. Recent evidence for a membrane site of action of estrogen has been reported for uterine cells and an early intracellular accumulation of calcium has been clearly demonstrated following estrogen administration [30]. The physiological implication of such early effects of estradiol on the passive membrane properties and excitability of pituitary cells remains unclear. In the case of TRH, which exerts both a short-term effect on PRL release and a long-term effect on PRL synthesis, the TRH-induced electrical activity may be associated with the stimulation of PRL release. 17β-E has been shown to have a long-term effect on PRL synthesis. Nevertheless, the possibility that PRL release is stimulated within minutes after 17β-E injection, which would be consistent with the Ca^{2+}-dependent electrical effect, has not yet been investigated. It is also possible that Ca^{2+} is required to initiate the long-term effect of 17β-E on PRL synthesis. These changes in ionic permeability of the membrane may also initiate other intracellular processes triggered by the steroid such as growth stimulation and cell division.

Effects of the Prolactin-Inhibiting Factors Dopamine and GABA

PRL-secreting cells are, at least partially, under the stimulatory influence of TRH and of estrogen. Nevertheless, they are mainly controlled by inhibitory substances among which dopamine is one of the most potent, *in vivo* as well as *in vitro*. In the GH3/B6 cell line, DA ejected in the close vicinity of a cell provoked an inhibition of firing within 30–60 sec (fig. 3). Similar results were observed in teleostean prolactin cells [36]. Under our experimental conditions [11, 12], this inhibitory effect is concomitant with a rapid decrease of the input resistance without any detectable change in the resting membrane polarization (fig. 3). In addition, DA inhibits the TRH-induced action potentials. Conversely, the DA antagonists haloperidol ($10^{-6}M$) and chlorpromazine ($10^{-6}M$) suppress the inhibitory effect of DA on action potentials when they are administered to a spontaneously firing cell. The DA inhibition of both spontaneous and TRH-induced action potentials may be related to the effect of DA on PRL secretion. It has been previously shown that the DA agonist CB 154 decreased TRH-induced PRL release in GH3/B6 cells without modifying the binding of ^{3}H-TRH.

The fact that DA inhibits both the firing and hormone release supports the hypothesis that Ca^{2+}-dependent action potentials are involved in the

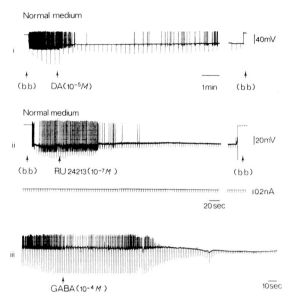

Fig. 3. Effects of DA, DA agonist, and GABA on the firing of action potentials in the GH3/B6 PRL-secreting cell line. As we have previously demonstrated that estrogen antagonized the action of DA [12], cells were cultured for 24 h in an estrogen-free medium. DA and RU 24213 (DA agonist), obtained respectively from Hoffman-La Roche and Roussel UCLAF, were dissolved in bathing solution. DA and RU 24213 depressed the firing of spontaneous action potentials and reduced the membrane resistance. Vertical bars represent a measure of the membrane resistance obtained by injecting a current of 0.16 nA with a bridge amplifier. DA had no effect when Haloperidol (Lebrun) or Chlorpromazine (Specia) were ejected before the administration of DA on the cell being recorded. GABA also inhibited the action potential firing and decreased the membrane resistance.

mechanism of release of anterior pituitary hormones [35, 36]. The mechanism by which DA inhibits the firing of PRL cells is not clearly understood at present. From our experiments, this mechanism appears to be different from that already reported for central neurones [18], in that no change in the membrane polarization was observed. The decrease in excitability induced by DA may be explained by depression of the spike-triggering mechanism resulting from an elevation of the spike threshold. Such voltage-dependent conductance effects have been termed 'neurohormonal' by *Barker et al.* [2] so as to distinguish them from 'neurotransmitter' effects which are voltage-independent. For example, the epinephrine regulation of spike conductance in cardiac muscle and the norepinephrine depression of Ca^{2+}

spike conductance in synaptic neurons [21] illustrate 'neuro-hormonal' actions of catecholamines on voltage-dependent conductances.

Finally, we observed that GABA was also able to inhibit the spontaneous firing activity of GH3/B6 cells (fig. 3). This inhibitory effect of GABA was apparently achieved by a decrease in the membrane resistance. Moreover, GABA decreases the size of the Ca^{2+}-dependent action potentials. This observation is consistent with various studies showing a direct inhibitory effect of GABA on the secretion of PRL by pituitary cells [33, 34]. As in the case of DA, the exact mechanism of this inhibition remains to be determined.

Conclusion

The neuroendocrine mechanisms which govern PRL secretion are complex. A number of substances are involved; they act either directly at the pituitary level or indirectly through a hypothalamic step.

It has been show that TRH is a physiological stimulatory factor for PRL secretion in a wide range of species including rat and man. The stimulatory effect of TRH on PRL secretion has also been shown *in vitro* on prolactin-secreting clonal cell lines [13, 15, 16, 29, 38] and in normal primary cultures of rat pituitary [42].

Ostlund et al. [29] have demonstrated that the acute release of PRL and growth hormone induced by TRH in GH3 is similar to that provoked by the addition of KCl to the bathing medium. Both the TRH and potassium-induced acute hormone release are calcium-dependent and are suppressed by EGTA, a calcium chelator, or by substituting magnesium for calcium. In our investigations, the application of high K^+ concentrations resulted in depolarization of the cell membrane which was then able to trigger Ca^{2+}-dependent action potentials.

In addition, these cells have been shown to respond to TRH with a prolonged discharge of calcium-dependent action potentials. The time course of the electrical response to TRH is consistent with that of the acute hormone release in response to TRH and KCl. Like the electrical effect, the hormone release can be demonstrated about 2 min after the stimulation by KCl or TRH. Thus, the response of acute hormone release and the electrical response appear to be closely associated and calcium provides the common key for both. We propose that action potentials provide the calcium influx which in turn is necessary to stimulate hormone release. Such a hypothesis is in agreement with the stimulus-secretion coupling theory proposed by *Douglas* [9].

The mechanism by which TRH facilitates calcium entry through firing of action potentials still remains to be elucidated. However, it should be noted that TRH induces action potential firing without any noticeable depolarization; the action of this substance differs therefore from that of high K^+. This observation is consistent with previous works by *Martin et al.* [22] showing that, at the pituitary level, depolarization is not tightly linked to the release process. The mechanism of action of TRH on pituitary cells differs also from that reported for spinal neurons since, in the latter model, the increase in excitability was observed to be related to depolarization [28].

The relationship between sex steroids and PRL is well documented [8]. Estrogen induces a preovulatory surge of PRL in rat [24] and influences PRL levels in monkey [25, 32] and human [43]. It has been shown that estrogen may act directly on the pituitary and also through a hypothalamic action. Estradiol stimulates PRL production in GH3 cells [16]. This stimulatory effect was significant after 4 days and reached a maximum by 10 days. Furthermore, 17β-E increased the number of membrane receptors for TRH [14, 15] and stimulated both basal and PRL response to TRH in different systems [15, 41, 44]. However, the short-term effect of 17β-E on the electrical activity of GH3/B6 cells is hardly compatible with the long-term effect of the steroid on PRL production. A possible stimulation of PRL release occurring within minutes following 17β-E administration, which would be consistent with the Ca^{2+} dependent effect induced by direct application of 17β-E, has not yet been investigated. Also, it is possible that the facilitation of Ca^{2+} entry is required for some other intracellular process triggered by the steroid. It is therefore worthwhile to note that the early effect of estradiol on calcium spiking activity which we observed is consistent with previous observations by *Pietras and Szego* [30] showing a rapid uptake of Ca^{2+} following administration of estrogens to isolated endometrial cells. These observations strongly suggest a membrane site of action for the steroid [30, 31].

As could be expected from the stimulus-secretion coupling hypothesis [9], DA inhibits action potential firing in prolactin secreting pituitary cells. This inhibitory effect does not seem to be mediated by a change in membrane polarization and differs therefore from the mechanisms of action of DA at the central level [18]. Also, GABA inhibits Ca^{2+}-dependent action potentials in prolactin secreting cells. These results are in agreement with the inhibitory effect of GABA on prolactin secretion [33, 34] and suggest a physiological role for this substance directly on pituitary cells. As for DA, the exact mechanism of action of GABA in inhibiting Ca^{2+}-dependent action potentials in GH3/B6 is not yet known but it certainly differs from its action on neurons as a neurotransmitter.

References

1 Barker, J.L. and Smith, T.G.: Peptides as neurohormones; in Cowan and Ferrendelli. Biological approaches to neurons. Neuroscience Symp., 1977, vol. II, pp. 340–373,
2 Barker, J.L.; Gruol, D.L.; Huang, L.M.; Neale, J.H., and Smith, T.G.: Enkephalin: pharmacologic evidence for diverse functional roles in the nervous system using primary cultures of dissociated spinal neurons; in Terenius, Characteristics and function of opioids (Elsevier/North-Holland, Amsterdam, in press).
3 Bauer, K.: Degradation of thyroliberin: possible physiological role of neuropeptide degrading enzymes; in Vincent and Kordon, Cell biology of hypothalamic neurosecretion, pp. 477–486 (CNRS, Paris 1978).
4 Bauer, K.; Graf, K.J.; Faivre-Bauman, A.; Beier, S.; Tixier-Vidal, A., and Kleinkauf, H.: Inhibition of prolactin secretion by histidyl-proline diketopiperazine. Nature, Lond. *274:* 174–175, (1978).
5 Biales, B.; Dichter, M., and Tischler, A.: Electrical excitability of cultured adrenal chromaffin cells. J. Physiol., Lond. *262:* 743–753 (1976).
6 Brandt, B.L.; Hagiwara, S.; Kidokoro, Y., and Miyazaki, S.: Action potentials in the rat chromaffin cell and effects of acetylcholine. J. Physiol., Lond. *263:* 417–439 (1976).
7 Chen, C.L. and Meites, J.: Effects of estrogen and progesterone on serum and pituitary prolactin levels in ovariectomized rats. Endocrinology *86:* 503–505 (1970).
8 Pozo, E. del and Brownell, J.: Prolactin. I. Mechanisms of control, peripheral actions and modification by drugs. Hormone Res. *10:* 143–172 (1979).
9 Douglas, W.W.: Stimulus-secretion coupling: the concept and clues from chromaffin and other cells. Br. J. Pharmacol. *34:* 451–474 (1968).
10 Douglas, W.W. and Taraskewich, P.S.: Action potentials in gland cells of rat pituitary pars intermedia: inhibition by dopamine, an inhibitor of MSH secretion. J. Physiol., Lond. *285:* 171–184 (1978).
11 Dufy, B.; Vincent, J.D.; Fleury, H.; Du Pasquier, P.; Gourdji, J., and Tixier-Vidal, A.: Membrane effects of thyrotropin-releasing hormone and estrogen shown by intracellular recording from pituitary cells. Science *204:* 509–511 (1979).
12 Dufy, B.; Vincent, J.D.; Fleury, H.; Du Pasquier, P.; Gourdji, J., and Tixier-Vidal, A.: Dopamine inhibition of action potentials in a prolactin secreting cell line is modulated by oestrogen. Nature, Lond. *282:* 855–857 (1979).
13 Gautvik, K.M. and Tashjian, A.H.: Effects of cations and colchicine on the release of prolactin and growth hormone by functional pituitary tumor cells in culture. Endocrinology *93:* 793–799 (1973).
14 Gershengorn, M.C.; Marcus Samuels, B.E., and Geras, E.: Estrogens increase the number of thyrotropin releasing hormone receptors on mammotropic cells in culture. Endocrinology *105:* 171–176 (1979).

15 Gourdji, D.: Characterization of thyroliberin (TRH) binding sites and coupling with prolactin and GH secretion in rat pituitary cell lines; in McKerns and Jutisz, Synthesis and release of adenohypophyseal hormones (Plenum Press, New York, in press).

16 Haug, E. and Gautvik, K.M.: Effects of sex steroids on prolactin secreting rat pituitary cells in culture. Endocrinology 99: 1982–1989 (1976).

17 Kidokoro, Y.: Spontaneous calcium action potentials in a clonal pituitary cell line and their relationship to prolactin secretion. Nature, Lond. 258: 741–743 (1975).

18 Kitai, S.T.; Sugimori, M., and Kocsis, J.D.: Excitatory nature of dopamine in the nigro-caudate pathway. Exp. Brain Res. 24: 351–363 (1976).

19 Lieberman, M.E.; Maurer, R.A., and Gorski, J.: Estrogen control of prolactin synthesis in vitro. Proc. natn. Acad. Sci. USA 12: 5946–5949 (1978).

20 MacLeod, R.M.: Influence of dopamine, serotonin and their antagonists on prolactin secretion; in Hubinont, L'Hermite and Robyn, Prog. reprod. Biol., vol. 2., pp. 54–68 (Karger, Basel 1977).

21 McAfee, D.A. and Yarowsky, P.J.: Calcium dependent potentials in the mammalian sympathetic neurone. J. Physiol., Lond. 290: 507–523 (1979).

22 Martin, S.; York, D.H., and Kraicer, J.: Alterations in transmembrane potential of adenohypophysial cells in elevated potassium and calcium-free media. Endocrinology 92: 1084–1088 (1973).

23 Matthews, E.K. and Sakamoto, Y.: Electrical characteristics of pancreatic islet cells. J. Physiol., Lond. 246: 421–437 (1975).

24 Meites, J.; Lu, K.H.; Wuttke, W.; Welsch, C.W.; Nayasawa, H., and Quadri, S.K.: Recent studies on functions and control of prolactin secretion in rats. Recent Prog. Horm. Res. 28: 471–516 (1972).

25 Milmore, J.E.: Influence of ovarian hormones on prolactin release in the rhesus monkey. Biol. Reprod. 19: 593–596 (1978).

26 Moriarty, C.M.: Role of calcium in the regulation of adenohypophysial hormone release. Life Sci. 23: 185–194 (1978).

27 Morin, A.; Tixier-Vidal, A.; Gourdji, D.; Kerdelhue, B., and Grouselle, D.: Effects of thyreotrope-releasing hormone (TRH) on prolactin turnover in culture. Mol. cell. Endocrinol. 3: 351–373 (1975).

28 Nicoll, R.A.: Excitatory action of TRH on spinal motoneurones. Nature, Lond. 265: 242–243 (1977).

29 Ostlund, R.E.; Leung, J.T.; Hajek, S.W.; Winokur, T., and Melman, M.: Acute stimulated hormone release from cultured GH3 pituitary cells. Endocrinology 103: 1245–1252 (1978).

30 Pietras, R.J. and Szego, C.M.: Endometrial cell calcium and oestrogen action. Nature, Lond. 253: 357–359 (1975).

31 Pietras, R.J. and Szego, C.M.: Estrogen receptors in uterine plasma membrane. J. Steroid Biochem. 11: 1471–1483 (1979).

32 Quadri, S.K. and Spies, H.G.: Cyclic and diurnal patterns of serum prolactin in the rhesus monkey. Biol. Reprod. 14: 495–501 (1976).

33 Racagni, G.; Apud, J.A.; Locatelli, V.; Cocchi, D.; Nistico, G.; Di Giorgio, R.M., and Muller, E.E.: GABA of CNS origin in the rat anterior pituitary inhibits prolactin secretion. Nature, Lond. 281: 575–578 (1979).

34 Schally, A.V.; Redding, T.W.; Arimura, A.; Dupont, A., and Linthicum, G.L.:

Isolation of γ-aminobutyric acid from pig hypothalami and demonstration of its prolactin release-inhibition (PIF) *in vivo* and *in vitro*. Endocrinology *100:* 681–691 (1977).

35 Taraskevitch, P.S. and Douglas, W.W.: Action potentials occur in cells of the normal anterior pituitary gland and are stimulated by the hypophysiotropic peptides thyrotropin releasing hormone. Proc. natn. Acad. Sci. USA *74:* 4064–4067 (1977).

36 Taraskevitch, P.S. and Douglas, W.W.: Catecholamines of supposed inhibitory hypophysiotropic function suppress action potentials in prolactin cells. Nature, Lond. *276:* 832–834 (1978).

37 Tashjian, A.H.; Bancroft, F.C., and Levine, L.: Production of both prolactine and growth hormone by clonal strains of rat pituitary tumor cells. J. Cell Biol. *47:* 61–70 (1970).

38 Tashjian, A.H.; Barousky, N.J., and Jensen, D.K.: Thyrotropin releasing hormone: direct evidence for stimulation of prolactin production by pituitary cells in culture. Biochem. biophys. Res. Commun. *43:* 516–523 (1971).

39 Tashjian, A.H.; Barowsky, N.J., and Jensen, D.K.: Thyrotropin releasing hormone: direct evidence for stimulation of prolactin production by pituitary cells in culture. Biochem. biophys. Res. Commun. *43:* 516–523 (1971).

40 Tashjian, A.H.; Lomedico, M.E., and Maina, D.: Role of calcium in the thyrotropin-releasing-hormone stimulated release of prolactin from pituitary cells in culture. Biochem. biophys. Res. Commun. *81:* 798–806 (1978).

41 Tixier-Vidal, A.; Brunet, N., and Gourdji, D.: Morphological and molecular aspects of the regulation of prolactin secretion by rat pituitary cell lines; in Robyn and Harter, Progress in prolactin physiology and pathology, pp. 29–43 (Elsevier, New York 1978).

42 Vale, W.; Blackwell, R.; Grant, G., and Guillemin, R.: TRF and thyroid hormones on prolactin secretion by rat anterior pituitary cells *in vitro*. Endocrinology *93:* 26–33 (1973).

43 Vekemans, M.; Delvoye, P.; L'Hermite, M., and Robyn, C.: Serum prolactin levels during the menstrual cycle. J. clin. endocr. Metab. *44:* 989–993 (1977).

44 Vician, L.; Shupnik, M.A., and Gorski, J.: Effects of estrogen on primary ovine pituitary cell cultures. Stimulation of prolactin secretion, synthesis, and preprolactin messenger ribonuclei acid activity. Endocrinology *104:* 736–743 (1979).

B. Dufy, PhD, INSERM U 176, Rue Camille Saint-Saens, F-33077 Bordeaux-Cedex (France)

Discussion

Friesen: Have you ever had the chance to test in your system TRH analogues which are known to be inactive? For example, does the de-amidated version of TRH affect electrical potential?

Dufy: We do not know that yet.

Friesen: Do you have any information yet on the effective concentration of TRH in the fluid bathing the cell?

Dufy: The concentration was 2 µg/ml, and we injected between 2 and 5 nl directly onto the cell membrane.

Friesen: The attempt to correlate electrophysiological data with hormonal changes and secretion rates for these cells is very interesting. Do you think it possible that there is, even in the single cell-line, a certain heterogeneity, so that some cells are responding differently? In that vein, do you think that it would ever be possible to determine prolactin release from a single cell, along with electrophysiological data?

Dufy: That is very difficult, because with one cell there is very little secretion. We have done some work with pituitary adenomas which have been removed by surgery, also, and recently we had pituitary adenomas secreting large amounts of hormone, and I hope to be able, perhaps, in the near future to record levels not from one cell but from a cluster of, let us say, ten cells, and to record the electrical activity of ten cells.

Bern: In my laboratory, we use a fish model system. This is very convenient: one can cut out a lobe of the fish pituitary which contains almost entirely prolactin cells. We have also found the importance of calcium with regard to prolactin secretion from these cells and have been looking into the effect of somatostatin: this is a very effective inhibitor of prolactin secretion. I have not heard somatostatin mentioned as a possible candidate for prolactin inhibition in mammalian systems: does anyone have any comment?

Neill: I believe that somatostatin inhibits prolactin release in the GH3 cell.

Thorner: Most people's experience, certainly in the human, is that somatostatin infusions give a very poor effect on prolactin secretion. Some people say that there is a small inhibitory effect; others say that there is no effect at all. In the rat, using the same technique that I showed before – the continuous perfusion of isolated rat cells, normal anterior pituitary cells – somatostatin in high doses produces a very minimal effect, maybe 10 or 20% inhibition.

Willoughby: Has it been possible to alter the electrical activity of these cells by iodophoresing prolactin, or altering prolactin concentration in the medium?

Dufy: We have not done it yet.

Neill: Do I remember correctly that several groups have looked to find dopamine receptors in the GH3 cell and have been unable to find them? The point is, of course, that with these nice effects on the cell, one wonders why receptors have not been found.

Dufy: I had a lot of trouble finding an inhibitory effect of dopamine on these cells. I think that there is an interaction with oestrogen. The cells are cultured with fetal calf serum and a lot of other things, but oestrogen is in the recording medium. We get a very small effect on this culture, but when the oestrogen in the culture is depleted, we can find a good level of inhibition. Perhaps there is an interaction between oestrogen and dopamine. As I showed, there is certainly a difference in the mechanism of dopamine at the brain level and at the pituitary level. At the brain level, dopamine acts at postsynaptic and presynaptic levels on action potentials via both sodium and potassium channels. Here it is acting on calcium.

Evidence for the Existence of the Physiological Prolactin-Releasing Factor and its Possible Role

S.H. Shin and H.J. Chi

Department of Physiology, Abramsky Hall, Queen's University, Kingston, Ont., and Natural Products Research Institute, Seoul National University, Seoul

Hypothalamic prolactin inhibiting factor (PIF) plays an important role in the control of the basal secretion of prolactin from the adenohypophysis [2, 4, 9, 17]. The chemical identity of the PIF has not been definitively established, but the general consensus is that dopamine is the major, if not the sole, physiological PIF [16]. The increased secretion of prolactin induced by acute stress is believed to be mediated through prolactin-releasing factor (PRF) but this is without direct experimental evidence.

There is evidence that certain fractions derived from hypothalamic extracts can stimulate prolactin secretion. The only chemically characterized component of hypothalamic extracts which will stimulate prolactin secretion is thyrotropin-releasing hormone (TRH) [30], but TRH is an unlikely candidate for the major physiological role [23]. Plasma thyroid-stimulating hormone (TSH) was not elevated [23], but rather decreased [10] by an ether stress which elevated plasma prolactin to more than 3 times the normal concentration [23]. Also, TRH raised plasma TSH to more than 5 times the normal concentration, while the elevation of plasma prolactin concentration in the male rat was insignificant. The dissociation between the elevation of plasma TSH and prolactin concentration with ether stress and TRH injection indicates that TRH is not the major physiological PRF in the normal male rat [23]. However, it should be pointed out that TRH does elevate the circulating prolactin level 6-fold in the estradiol-primed rat. In this animal preparation, the prolactin concentration is already elevated to about 10 times the normal concentration by the estradiol implantation [27].

Identification of PRF activity in hypothalamic extracts does not necessarily mean that such activity will be physiologically significant and that it will be secreted into the hypophysial-portal vessels to stimulate prolactin secretion from the adenohypophysis. The releasing activity might be an intermediate product of biosynthesis or biodegradation, a breakdown product during the process of extraction or a local transmitter.

We have continued to investigate the possibility that, in acute stress, PRF may play an important role in stimulating prolactin secretion. Since we have characterized the temporal relationship between ether stress and prolactin secretion [6], the administration of ether vapour was chosen as the acute stress in this series of experiments.

Two different approaches were used to establish the existence of a physiological PRF. First, we attempted to mask changes in PIF secretion, and thus to remove its influence on prolactin secretion by infusing massive amounts of dopamine. We then exposed the rat to ether vapour, expecting that only PRF will be functional. The second approach was to block dopaminergic receptors by a 'specific' pharmacological blocking agent, pimozide, so that the dopaminergic PIF would not be functional and only the effect of PRF would be shown.

Materials and Methods

Prolactin secretion is sensitive to environmental stress [14, 31]. We used a special chamber equipped with a one-way glass for housing the animals during the blood sampling so that we could observe any unusual behavior of the experimental animals. Blood samples were taken through an atrial indwelling cannula [6]. In order to avoid possible stress by changes of fluid balance and to avoid possible artifacts due to infusion of suspended red blood cells, we did not return red blood cells, but instead we took a minimum volume of blood (70 μl) for assay of triplicate plasma samples. The total volume taken for 60 sequential samples was approximately 4.2 ml or approximately 15% of the total blood volume of a 350-gram rat over a period of 2 h. No sudden prolactin secretion was observed through the course of the taking of the 60 samples, indicating that a 15% loss of blood can be tolerated without serious effect on prolactin secretion [6, 25].

Dopamine-HCl (Sigma), Pimozide (a gift from Dr. *W.J. Forgiel*, McNeil Lab. Canada), ether (Mallinckrodt) and radioimmunoassay kit for prolactin (NIAMDD) were used in this study.

Results

Dopamine Infusion into the Hypothalamic-Hypophysial System

When unrestrained, freely moving rats, receiving dopamine by infusion through the right carotid artery at a rate of 30 ng/10 μl/min and passively

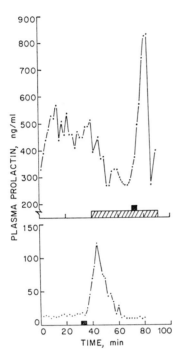

Fig. 1. Effect of ether stress on prolactin release in dopamine-infused rats. The normal male rat (lower panel), or estradiol-implanted male rat (upper panel) which was receiving dopamine by infusion through the carotid artery at a rate of 30 ng/10 μl/min, was exposed to ether vapour for 4 min as shown by the black bars on the abscissae. Dopamine was infused either throughout the experiment (lower panel) or 40 min after the beginning of the experiment as indicated by a hatched bar on the abscissa (upper panel). Circulating prolactin concentration was monitored every 2 min. Each datum point indicates the mean value of triplicate assays. Each panel represents an individual rat. Similar responses were seen in a total of 5 normal male rats and in 4 estradiol-primed rats.

donating blood samples from the right atrium every 2 min, were exposed to ether for 4 min, circulating prolactin was consistently elevated by the ether stress (fig. 1). Infusion of dopamine alone (30 ng/10 μl/min) did not cause a significant change in the basal prolactin level in the male rat. When an estradiol-packed Silastic capsule was implanted in a castrated male rat for 3 weeks, the circulating prolactin concentration was elevated by more than 10 times the normal circulating level of 10–20 ng/ml [26]. This elevated prolactin level was then significantly lowered by the infusion of dopamine at a rate of 30 ng/10 μl/min (fig. 1). When such estradiol-implanted rats were

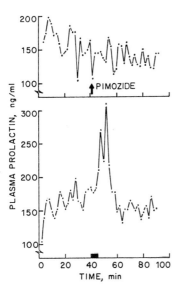

Fig. 2. Rats were treated with pimozide (3 mg/kg) 2 h before the experiment and blood samples were then collected every 2 min. 40 min after monitoring circulating plasma prolactin, rats were either injected with pimozide (300 µg/kg) as indicated by an arrow (upper panel) or exposed to ether vapour for 4 min as indicated by a black bar on the abscissa (lower panel). Each datum point indicates the mean value of triplicate assay. Each panel represents an individual rat. Similar responses to a second injection of pimozide were seen in a total of 3 rats and to the ether stress in 6 rats.

exposed to ether for 4 min during the dopamine infusion, a prolactin surge was observed resulting in a 3-fold elevation from the pre-ether stress concentration (fig. 1).

'Elimination' of the Role of Dopaminergic Receptors in the Hypothalamic-Hypophysial System

First, we established a dose-response relationship between pimozide and the plasma prolactin concentration [25]. There was a graded dose-response relationship between the amount of pimozide administered and the plasma prolactin concentration at the lower doses of pimozide (30, 60, and 300 µg/kg), but at the higher doses (300, 800, and 3,000 µg/kg), the maximum plasma prolactin concentration was not further increased. 300 µg/kg pimozide was the lowest dose producing the maximum prolactin secretion [25]. Therefore, it would be reasonable to assume that 10 times the dose eliciting the maximum response (3 mg/kg) will completely block all functioning dopaminergic PIF receptors.

Second, in order to positively demonstrate that all of the dopaminergic PIF receptors were still blocked by pimozide at the time of ether stress, animals that had received pimozide (3 mg/kg), 2 h and 40 min previously, received a second injection (300 µg/kg) or were subjected to 4 min of ether stress. There was no change in the circulating prolactin concentration following the second injection of pimozide (fig. 2), indicating that all of the available dopaminergic PIF receptors had likely been completely blocked by the first dose (3 mg/kg) of pimozide. In this situation, lacking functional dopaminergic receptors, there were pulsatile fluctuations in the circulating prolactin concentration throughout the experiment. These results suggest that the dopaminergic PIF cannot be participating actively in the generation of pulsatile prolactin secretion (fig. 2).

When rats were then exposed to an ether stress 2 h and 40 min after the pimozide injection, the prolactin concentration was dramatically elevated, even though there were no functional dopaminergic PIF receptors available (fig. 2). The elevation of prolactin concentration induced by ether stress was highly significant by the paired Student's t-test.

Discussion

Dopamine is probably the physiological PIF. Dopamine is found in the hypothalamus and is secreted into the hypophysial-portal system [1, 11, 20]. Dopamine agonists, such as apomorphine and ergot alkaloids, inhibit prolactin release [15, 18, 26] while the dopaminergic antagonist, pimozide, stimulates prolactin release [19, 25]. There are dopaminergic receptors in the adenohypophysis which can be stimulated (to inhibit prolactin secretion) by dopamine agonists and can be inhibited by dopamine antagonists [24]. These facts are consistent with the concept that dopamine or a dopamine agonist is the physiological PIF receptor through which tonic inhibition or prolactin secretion is exerted.

Several other putative neurotransmitter substances, such as γ-aminobutyric acid (GABA), are found in hypothalamic extracts [22, 29], but GABA is not likely a major physiological PIF since micromolar concentrations of GABA are required to inhibit prolactin release [8].

Since the existence of PRF activity in hypothalamic extracts is established [3, 7, 13, 28, 32], it is conceivable that the rapid surge of prolactin secretion induced by acute stress [6] might be due to the stimulation of PRF secretion rather than to an inhibition of PIF secretion. No attempt was made to prove the existence of the physiological PRF until *Valverde et al.* [33] endeavored to establish a role for PRF. They tried to deplete endogenous dopamine, a putative PIF [16], by a single subcutaneous injection of reserpine (500 μg/100 g b.w.) 1 day before the experiment, and then to stimulate prolactin secretion by ether stress. Since prolactin secretion in the catecholamine-depleted rat was stimulated by the ether treatment, *Valverde et al.* [33] concluded that the effect of ether stress on prolactin secretion was due to stimulation of secretion of PRF.

We have further pursued the problem in an attempt to answer the question as to whether prolactin secretion induced by acute stress is due to inhibition of PIF secretion or to stimulation of PRF secretion. If prolactin secretion induced by ether treatment is due to the inhibition of PIF secretion, then there should be no effect of ether stress on prolactin secretion when a high concentration of dopamine (3,000 ng/ml), a putative PIF, is constantly infused at a moderate perfusion rate throughout the experiment. Thus, a relatively high concentration of dopamine would be in the portal blood even though the hypothalamus might not be secreting dopamine (or a dopaminergic PIF, if not dopamine). The infusion of dopamine would result in a chronic inhibition of prolactin secretion. We have arbitrarily chosen an infusion rate of 10 μl/min, which is slightly lower than the stalk plasma flow rate (approximately 12–25 μl/min) [21], in order to avoid a drastic change in the normal composition of the portal blood. A dopamine infusion of 30 ng/10 μl/min through the carotid artery should be more than sufficient to offset possible suppression of endogenous PIF secretion from the hypothalamus since the dopamine concentration in the hypophysial portal blood in the male rat is not more than 1 ng/ml [1].

We could not demonstrate that dopamine depresses the basal circulating prolactin concentration in the normal male under our experimental conditions. This is most likely due to the fact that the normal circulating level of prolactin in the male rat is very low, implying that the rate of prolactin secretion is already at a nadir due to either high portal PIF and/or low portal PRF. It is not possible to establish directly how much of the infused dopamine is actually passing through the hypothalamic portal blood since at present it is technically impossible to obtain portal blood from an undisturbed and unanesthetized animal.

In order to demonstrate that dopamine infusion results in an effective dopamine level in the portal circulation, we began the experiment with estrogen-pretreated rats, since it is known that the implantation of estradiol into castrated rats will elevate prolactin levels [5, 12, 26]. We could then demonstrate an effective inhibition of this augmented prolactin level with dopamine infusion (fig. 1), thus indicating that a substantial amount of the administered dopamine must be passing through the hypophysial portal vessels (fig. 1). We then induced a prolactin surge by ether administration while maintaining the dopamine infusion (which would 'mask' a change in endogenous PIF release), indicating that it is not inhibition of PIF that gives rise to the prolactin surge induced by ether stress. Since the prolactin secretion induced by the ether stress is not due to the inhibition of dopaminergic PIF secretion, it must be due to the stimulation of PRF secretion.

In order to further support the role of the physiological PRF, we have completely blocked all the available dopaminergic receptors and then applied an acute ether stress. The plasma prolactin concentration was consistently elevated by the ether stress after completely 'eliminating' functional dopaminergic receptors by a supramaximal dose of pimozide. This is further supporting evidence that the hypothalamic dopaminergic PIF is not involved in stress-induced prolactin secretion.

Conclusion

Acute stress stimulates prolactin secretion and the effect of the stress on prolactin secretion must be mediated through hypothalamic secretion of PIF and/or PRF. We have eliminated the possibility that the prolactin surge is due to the inhibition of dopaminergic PIF secretion by two different approaches, infusion of excess dopamine and blockage of dopaminergic receptors. It was also shown during this series of experiments that the pulsatile secretion of prolactin is not mediated through inhibition of hypothalamic PIF secretion.

All the evidence that we have gathered is consistent with the concept that the basal level of prolactin secretion is maintained by PIF secretion but that the prolactin surge induced by acute stress is mediated through PRF secretion.

Acknowledgements

This work was supported by the Medical Research Council of Canada. The authors wish to thank Dr. *A.F. Parlow* of the NIAMDD Rat Pituitary Hormone Distribution

Program for the prolactin RIA kit, Dr. *J. Kraicer* and Dr. *A.E. Zimmerman* for stimulating discussion, Dr. *W.J. Forgeil*, McNeil Laboratory (Canada) Ltd. for pimozide and Miss *S. Ball* for excellent typing.

References

1 Ben-Jonathan, N.; Oliver, C.; Weiner, H.J.; Mical, R.S., and Porter, J.C.: Dopamine in hypophysial portal plasma of the rat during the estrous cycle and throughout pregnancy. Endocrinology *100:* 452–458 (1977).
2 Bishop, W.; Krulich, L.; Fawcett, C.P., and McCann, S.M.: The effect of median eminence (ME) lesions on plasma level of FSH, LH and prolactin in the rat. Proc. Soc. exp. Biol. Med. *136:* 925–927 (1971).
3 Boyd, A.E., III; Spencer, E.; Jackson, I.M.D., and Reichlin, S.: Prolactin releasing factor (PRF) in porcine hypothalamic extract distinct from TRH. Endocrinology *99:* 861–871 (1976).
4 Chen, C.L.; Amenomori, Y.; Lu, K.H.; Voogt, J.L., and Meites, J.: Serum prolactin levels in rat with pituitary transplants or hypothalamic lesions. Neuroendocrinology *6:* 220–227 (1970).
5 Chen, C.L. and Meites, J: Effects of estrogen and progesterone on serum and pituitary levels in ovariectomized rats. Endocrinology *86:* 503–505 (1970).
6 Chi, H.J. and Shin, S.H.: The effect of exposure to ether on prolactin secretion and the half-life of endogenous prolactin in normal and castrated male rats. Neuroendocrinology *26:* 193–201 (1978).
7 Dular, R.; LaBella, F.; Vivian, S., and Eddie, L.: Purification of prolactin releasing and inhibiting factors from beef. Endocrinology *94:* 563–567 (1974).
8 Enjalbert, A.; Ruberg, M.; Arancibia, S.; Fiore, L.; Priam, M., and Kordon, C.: Independent inhibition of prolactin secretion by dopamine and γ-aminobutyric acid *in vitro*. Endocrinology *105:* 823–826 (1979).
9 Everett, J.W.: Luteotropic function of autografts of the rat hypophysis. Endocrinology *54:* 685–690 (1954).
10 Fortier, C.; Delgado, A.; Ducomun, P.; Ducommun, S.; Dupont, A.; Jobin, M.; Kraicer, J.; MacIntosh-Hardt, B.; Marceau, H.; Mialhe, P.; Mialhe-Voloss, C.; Rerup, C., and Van Rees, P.: Functional interrelationships between the adenohypophysis, thyroid, adrenal cortex and gonads. Can. med. Ass. J. *103:* 864–874 (1970).
11 Gudelsky, G.A. and Porter, J.C.: Release of newly synthesized dopamine into the hypophysial portal vasculature of the rat. Endocrinology *104:* 583–587 (1979).
12 Kalra, P.S.; Fawcett, C.P.; Krulich, L., and McCann, S.M.: The effects of gonadal steroids on plasma gonadotropins and prolactin in the rat. Endocrinology *92:* 1256–1268 (1973).
13 Kokubu, T.; Sawano, S.; Shiraki, M.; Yamasaki, M., and Ishizuka, Y.: Extraction and partial purification of prolactin-release stimulating factor in bovine hypothalamus. Endocr. jap. *22:* 213–217 (1975).
14 Krulich, L.; Hefco, E.; Illner, P., and Read, C.B.: The effects of acute stress on the secretion of LH, FSH, prolactin and GH in the normal male rat, with comments on their statistical evaluation. Neuroendocrinology *16:* 293–311 (1974).

15 Lu, K.H.; Koch, Y., and Meites, J.: Direct inhibition by ergocornine of pituitary prolactin release. Endocrinology *89:* 229–233 (1971).

16 MacLeod, R.M.: Regulation of prolactin secretion; in Martini and Ganong, Frontiers in neuroendocrinology, vol. 4, pp. 169–194 (Raven Press, New York 1976).

17 Meites, J. and Nicoll, C.S.: Adenohypophysis: prolactin. A. Rev. Physiol. *28:* 57–88 (1966).

18 Ojeda, S.R.; Harms, P.G., and McCann, S.M.: Possible role of cyclic AMP and prostaglandin E_1 in the dopaminergic control of prolactin release. Endocrinology *95:* 1694–1703 (1974).

19 Ojeda, S.R.; Harms, P.G., and McCann, S.M.: Effects of blockade of dopaminergic receptors on prolactin and LH release: median eminence and pituitary sites of action. Endocrinology *94:* 1650–1657 (1974).

20 Plotsky, P.M.; Gibbs, D.M., and Neill, J.D.: Liquid chromatographic-electrochemical measurement of dopamine in hypophysial stalk blood of rats. Endocrinology *102:* 1887–1894 (1978).

21 Sarkar, D.K.; Chiappa, S.A.; Fink, G., and Sherwood, N.M.: Gonadotropin-releasing hormone surge in pro-estrous rats. Nature, Lond. *264:* 461–463 (1976).

22 Schally, A.V.; Redding, T.W.; Arimura, R.A.; Dupont, A., and Linthicum, G.L.: Isolation of gamma-aminobutyric acid from pig hypothalami and demonstration of its prolactin release-inhibiting (PIF) activity *in vivo* and *in vitro*. Endocrinology *100:* 681–691 (1977).

23 Shin, S.H.: Thyrotropin releasing hormone (TRH) is not the physiological prolactin releasing factor (PRF) in the male rat. Life Sci. *23:* 1813–1818 (1978).

24 Shin, S.H.: Dopamine induced inhibition of prolactin release from cultured adenohypophysial cells: spare receptors for dopamine. Life Sci. *22:* 67–74 (1978).

25 Shin, S.H.: Pulsatile secretion of prolactin in the male rat after pimozide administration is not due to pulsatile inhibition of PIF secretion. Life Sci. *24:* 1751–1756 (1979).

26 Shin, S.H. and Chi, H.J.: Unsuppressed prolactin secretion in the male rat is pulsatile. Neuroendocrinology *28:* 73–81 (1979).

27 Shin, S.H. and Piercy, M.: Estradiol induces TRH receptors to stimulate prolactin secretion. Proc. 6th Int. Congr. Endocrinology, Melbourne 1980, abstr.

28 Szabo, M. and Frohman, L.A.: Dissociation of prolactin-releasing activity from thyrotropin-releasing hormone in porcine stalk median eminence. Endocrinology *98:* 1451–1459 (1976).

29 Tappaz, M.L.; Brownstein, M.J., and Kopin, I.J.: Glutamate decarboxylase (GAD) and γ-aminobutyric acid (GABA) in discrete nuclei of hypothalamus and substantia nigra. Brain Res. *125:* 109–121 (1977).

30 Tashjian, A.H., Jr.; Barowsky, N.J., and Jensen, D.K.: Thyrotropin-releasing hormone: direct evidence for stimulation of prolactin production by pituitary cells in culture. Biochem. biophys. Res. Commun. *43:* 516–523 (1971).

31 Terry, L.C.; Saunders, A.; Audet, J.; Willoughby, J.O.; Brazeau, P., and Martin, J.B.: Physiological secretion of growth hormone and prolactin in male and female rats. Clin. Endocr. *6:* 19s–28s (1977).

32 Valverde-R., C.; Chieffo, V., and Reichlin, S.: Prolactin-releasing factor in porcine and rat hypothalamic tissue. Endocrinology *91:* 982–993 (1972).

33 Valverde-R., C.; Chieffo, V., and Reichlin, S.: Failure of reserpine to block ether-induced release of prolactin: physiological evidence that stress induced prolactin release is not caused by acute inhibition of PIF secretion. Life Sci. *12:* 327–335 (1973).

S.H. Shin, PhD, Department of Physiology, Abramsky Hall, Queen's University, Kingston, Ontario K7L 3N6 (Canada)

Discussion

Voogt: The interpretation was that if dopamine receptors have been completely blocked, the increase in prolactin observed with suckling must indicate a PRF. It could also indicate another PIF that is not dopamine activity but some other kind of factor, or that the rise is due to a decrease in that, rather than an increase in a PRF.

Shin: I expected that kind of issue to come out. I must admit that that is the weakest link of the chain. It is certainly possible.

Neill: The reason that some of us think that there is another PIF is that, first of all, some of the studies that I have done, in which we infused dopamine in physiological doses, show that we cannot fully suppress prolactin release. There seems to be a need to invoke a second PIF to explain our own observations. Then, there are the multiple observations in the literature which show very clear evidence for a PIF which is not dopamine. The case is very strong for there being a second PIF, not only from the quantitative studies that we have done, but because these activities have been demonstrated.

Shin: I fully agree with *Neill*, but when we look into the effects of the extract, we cannot exclude the idea that the effect is due to an intermediate product of biosynthesis, or biodegradation, or local neurotransmitter, and so on. We must be very careful when we deal with extracts, for artifacts can be easily produced and the active materials may not really be endogenous physiological control factors.

Effect of a Synthetic Enkephalin Analogue on Prolactin Secretion in Man and Rats

Yuzuru Kato, Nobrio Matsushita, Hideki Katakami and Hiroo Imura

Second Medical Clinic, Department of Medicine, Kyoto University Faculty of Medicine, Kyoto

Recent studies have revealed that neuropeptides present in the central nervous system influence the secretion of prolactin (PRL). We have previously reported that substance P [4], neurotensin [5], β-endorphin [6], α-endorphin [6], Met[5]-enkephalin [6] and vasoactive intestinal polypeptide (VIP) [7] stimulate PRL release in the rat.

In the present study, we examined the effect of a synthetic enkephalin analogue FK33-824 [9], Tyr-DAla-Gly-MePhe-Met(0)-ol, on plasma PRL levels in man and rats.

Materials and Methods

In animal experiments, Wistar strain male rats (Japan Animal Co., Osaka) were used throughout the experiments. They were anesthetized with urethane (150 mg/100 g body wt, i.p.) or kept unanesthetized with intraatrial catheters inserted through the jugular vein 2 weeks before the experiment. FK33-824 (Sandoz, Basel), β-endorphin (Daiichi Pharmaceutical Co., Tokyo) and naloxone (Endo Labs., New York) were injected intravenously or intraventricularly, and blood samples were withdrawn immediately before and 10, 20 and 40 min after the injection. The effect of TRH (Tanabe Pharmaceutical Co., Osaka), dopamine, FK33-824 and β-endorphin on PRL release from cultured anterior pituitary cells were also studied *in vitro* using the method described previously [5].

In human studies, FK33-824 was injected intramuscularly in normal subjects and patients with pituitary disorders. They were reexamined after a week with FK33-824 injection, which was followed by the administration of naloxone infused intravenously for 2 h. Blood samples were collected, through an indwelling needle inserted into a cubital vein, at 30 min intervals for 3 h.

Plasma human PRL and rat PRL levels were measured by specific radioimmunoassay, as previously reported [2, 3]. Student's t-test was used for statistical evaluation.

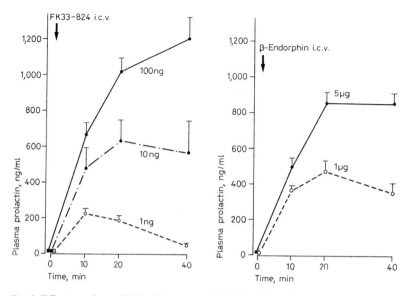

Fig. 1. Effect on plasma PRL of intraventricular injection of FK33-824 (1 ng, 10 ng and 100 ng per rat) and synthetic β-endorphin (1 μg and 5 μg per rat) in urethane-anesthetized rats. In this and subsequent figures, all values are the mean ± SE.

Results

Effect of FK33-824 on PRL Release in the Rat

Intraventricular injection of FK33-824 (1, 10, and 100 ng per rat) and β-endorphin (1 μg and 5 μg per rat) resulted in a significant and dose-related increase in plasma PRL levels in the urethane-anesthetized rat (fig. 1). Considering molar concentration, FK33-824 is 1,400 times more potent than β-endorphin in stimulating PRL release.

As shown in figure 2, intravenous injection of FK33-824 (10 μg/100 g body wt) also caused a significant increase in plasma PRL levels both in anesthetized rats and in unanesthetized, freely moving rats. Intravenous administration of naloxone (125 μg/100 g body wt), a specific opiate receptor blockade, significantly blunted plasma PRL increase induced by FK33-824.

Pretreatment with reserpine (1 mg/100 g body wt, i.p.) inhibited PRL response to FK33-824 in rats with or without anesthesia (fig. 2). The rise in plasma PRL induced by FK33-824 was also blunted by pretreatment with pimozide (50 μg/100 g body wt, i.p.), which was given 30 min earlier (fig. 3).

PRL release from cultured anterior pituitary cells was significantly

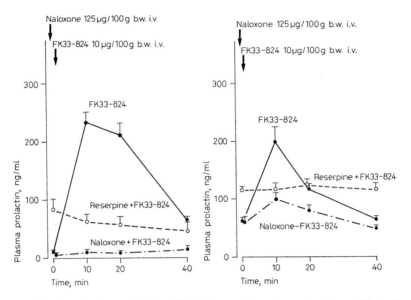

Fig. 2. Effect of naloxone (125 μg/100 g body wt, i.v.) and reserpine (1 mg/100 g body wt, i.p.) on PRL responses to intravenous injection of FK33-824 (10 μg/100 g body wt) in urethane-anesthetized rats (left panel) and in unanesthetized, freely moving rats with the cannula chronically inserted into the right atrium (right panel). Naloxone was injected 3 min before FK33-824 administration and reserpine was injected 20 h earlier.

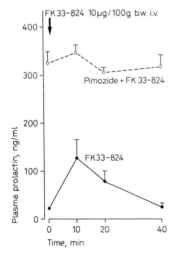

Fig. 3. Effect of pimozide (50 μg/100 g body wt, i.p.) on PRL release induced by FK33-824 in urethane-anesthetized rats. Pimozide was injected 30 min before the experiment.

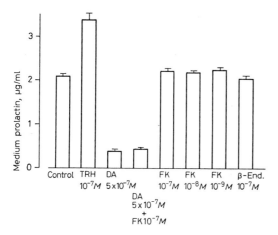

Fig. 4. Effect of TRH, dopamine, FK33-824 and β-endorphin on PRL release from cultured rat anterior pituitary cells. 5×10^6 cells were preincubated for 2 h in culture medium and then for 4 h in control, TRH ($10^{-7}M$), dopamine ($5 \times 10^{-7}M$), FK33-824 (10^{-7}–$10^{-9}M$), β-endorphin ($10^{-7}M$) and dopamine plus FK33-824 containing medium.

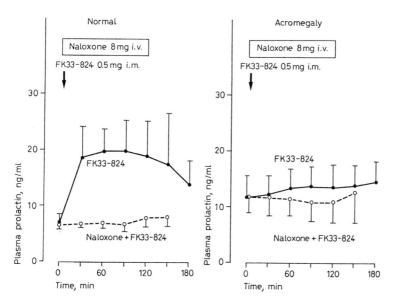

Fig. 5. Plasma human PRL levels following the intramuscular injection of FK33-824 (0.5 mg), and effect of naloxone (8 mg) infused intravenously on human PRL response to FK33-824 in 5 normal subjects (left panel) and 7 patients with acromegaly (right panel).

stimulated by TRH ($10^{-7} M$) and inhibited by dopamine ($5 \times 10^{-7} M$). Addition of FK33-824 (10^{-7} $10^{-9} M$) to the incubation medium had no significant effect by itself and did not attenuate the inhibition of PRL release induced by dopamine (fig. 4).

Effect of FK33-824 on PRL Release in Man

Intramuscular injection of FK33-824 (0.5 mg) caused a significant increase in plasma PRL levels in all normal subjects examined (fig. 5). Mean ($\pm$ SE) peak values of plasma PRL were 21.9 ± 5.9 ng ml, which were obtained 60–120 min after the injection. Concomitant intravenous infusion of naloxone (8 mg) significantly suppressed the plasma PRL response to the enkephalin analogue in all of these subjects.

In 5 out of 6 patients with acromegaly, FK33-824 did not stimulate PRL release as shown in figure 5. FK33-824 did not cause any increase in plasma PRL levels in patients with isolated PRL deficiency and patients after hypophysectomy. In a patient with Cushing's syndrome, plasma PRL increase induced by FK33-824 was exaggerated after adrenalectomy.

Discussion

Following the recent identification of neuropeptides with opiate like activity, it has been suggested that endogenous endorphins could be involved in the control of PRL secretion [1]. We demonstrated in the present experiments that a synthetic analogue of Met^5-enkephalin, FK33-824, stimulated PRL secretion in the rat. FK33-824 was more potent than β-endorphin in stimulating PRL secretion when the agents were injected intraventricularly. FK33-824 exerts a long-lasting and potent analgesic effect in animals after systemic and enteral administration [9]. We demonstrated that intravenous injection of FK33-824 caused a significant increase in plasma PRL in the rat regardless of anesthesia.

Specificity of the stimulating effect of FK33-824 on PRL release was demonstrated with naloxone, a specific opiate antagonist. Pretreatment with either reserpine, a depletor of catecholamines and serotonin, or pimozide, a specific inhibitor of dopaminergic activity, inhibited PRL response to FK33-824 in the rat, in which basal PRL levels were elevated. We observed in *in vitro* experiments that FK33-824 did not stimulate PRL release from cultured pituitary cells nor did it attenuate the inhibiting action of dopamine.

These results suggest that FK33-824 stimulates PRL secretion, possibly acting through the opiate receptor in the central nervous system, and that brain amines, at least dopamine, may be mediated in the action of the opioid peptide. These findings are in agreement with the hypothesis that presynaptic inhibiting opiate receptors are present on dopaminergic neurons of the tuberoinfundibular system [9].

Spampinato et al. [10] have recently reported a possible mediation of PRL release induced by enkephalins via the brain serotonin system. The possibility is not completely excluded from the present experiments that opioid peptides exert at least part of their hormonal effects through activation of serotoninergic mechanisms and that presynaptic stimulating opiate receptors are present on serotoninergic nerve endings in contact with the tuberoinfundibular dopaminergic system [1].

We observed that a single intramuscular injection of FK33-824 caused a significant increase in plasma human PRL in normal subjects. These results are in agreement with the previous reports [8, 11]. We further demonstrated that FK33-824 did not stimulate PRL release in patients with acromegaly. The mechanism remains to be further studied. Intravenous injection of TRH (500 μg) caused a significant increase in plasma PRL in these patients. It is possible, therefore, that the altered opiate receptor mechanism in the central nervous system might be involved. The action of FK33-824 may be influenced by adrenal function, since adrenalectomy caused an exaggerated increase in plasma PRL in a patient with Cushing's syndrome.

Acknowledgements

These studies were supported in part by grants from the Ministry of Education and the Ministry of Health and Welfare, Japan. We are indebted to the National Institute of Arthritis, Metabolism and Digestive Diseases for supplying materials for the human and rat PRL radioimmunoassays. We would like to thank Sandoz Ltd (Basel) and Endo Labs (New York) for the gift of FK33-824 and naloxone, respectively.

References

1 Dupont, A.; Cusan, L.; Ferland, L.; Lemay, A., and Labrie, F.: Evidence for a role of endorphins in the control of prolactin secretion; in Collu, Central nervous system effects of hypothalamic hormones and other peptides, pp. 283–300 (Raven Press, New York 1979).
2 Kato, Y.; Nakai, Y.; Imura, H.; Chihara, K., and Ohgo, S.: Effect of 5-hydroxytryptophan (5-HTP) on plasma prolactin levels in man. J. clin. Endocr. Metab. *38:* 695–697 (1974).

3   Kato, Y.; Chihara, K.; Ohgo, S., and Imura, H.: Effect of nicotine on the secretion of growth hormone and prolactin in rats. Neuroendocrinology *16:* 237–242 (1974).
4   Kato, Y.; Chihara, K.; Ohgo, S.; Iwasaki, Y.; Abe, H., and Imura, H.: Growth hormone and prolactin release by substance P in rats. Life Sci. *19:* 441–446 (1976).
5   Kato, Y.; Iwasaki, Y.; Abe, H.; Imura, H., and Yanaihara, N.: Effects of substance P, neurotensin, endorphins and vasoactive intestinal polypeptide (VIP) on plasma prolactin and growth hormone levels in rats. Proc. Symp. Chem. Physiol. Pathol., vol. 17, pp. 71–75 (1977).
6   Kato, Y.; Iwasaki, Y.; Abe, H.; Ohgo, S., and Imura, H.: Effects of endorphins on prolactin and growth hormone secretion in rats. Proc. Soc. exp. Biol. Med. *158:* 431–436 (1978).
7   Kato, Y.; Iwasaki, Y.; Iwasaki, J.; Abe, H.; Yanaihara, N., and Imura, H.: Prolactin release by vasoactive intestinal polypeptide in rats. Endocrinology *103:* 554–558 (1978).
8   Graffenried, B. von; del Pozo, E.; Roubicek, J.; Krebs, E.; Pöldinger, W.; Burmeister, P., and Kerp, L.: Effects of the synthetic enkephalin analogue FK33-824 in man. Nature, Lond. *272:* 729–730 (1978).
9   Roemer, D.; Buescher, H.H.; Hill, R.C.; Press, J.; Bauer, W.; Cardinaux, F.; Closse, A.; Hauser, D., and Huguenin, R.: A synthetic enkephalin analogue with prolonged parenteral and oral analgesic activity. Nature, Lond. *268:* 547–549 (1977).
10 Spampinato, S.; Locatelli, V.; Cocchi, D.; Vicentini, L.; Bajusz, S.; Ferri, S., and Müller, E.E.: Involvement of brain serotonin in the prolactin releasing effect of opioid peptides. Endocrinology *105:* 163–170 (1979).
11 Stubbs, W.A.; Delitala, G.; Jones, A.; Jeffcoate, W.J.; Edwards, C.R.W.; Ratter, S.J., and Besser, G.M.: Hormonal and metabolic responses to an enkephalin analogue in normal man. Lancet *ii:* 1225–1227 (1978).

Y. Kato, MD, Second Medical Clinic, Department of Medicine, Kyoto University Faculty of Medicine, 54 Kawahara-cho, Shogoin, Sakyo-ku, Kyoto (Japan)

Discussion

Veldhuis: As to the role of opiates in the control of prolactin secretion, there seems to be some important differences between species, at least in studies which have used naloxone. Perhaps I could mention, briefly, our own work using pharmacologic doses of naloxone – 0.2 mg/kg – in the normal human male and female: this really fails to exert any effect on basal prolactin concentration in peripheral plasma.

Kato: We had similar results with naloxone in normal subjects and patients with hyperprolactinaemia. In humans, I think that the basal prolactin levels are not regulated by the opioid peptide, but some stimulus, such as stress or anxiety, may stimulate prolactin secretion, mediates by the endogenous opioid peptide. But I want to say that the endogenous opioid peptide may have a role in abnormal growth hormone secretion in a part of patients with acromegaly. Usually, naloxone infusion did not change the plasma growth hormone levels, but we did find in 2 of our 12 acromegalic patients that naloxone inhibited plasma growth hormone levels. This effect was significantly attenuated by the injection of FK33-824.

Nyctohemeral Profiles of Plasma Prolactin after Jet Lag in Normal Man[1]

Daniel Désir, Marc L'Hermite, Eve Van Cauter, Anne Caufriez, Claude Jadot, Jean-Paul Spire, Pierre Noël, Samuel Refetoff, Georges Copinschi and Claude Robyn

Free University of Brussels, Brussels, and University of Chicago, Chicago, Ill.

Prolactin (PRL) is secreted episodically in normal man [1]. The nyctohemeral secretory profile is further characterized by a minor afternoon rise and a major nighttime rise in plasma concentrations of the hormone [2]. Daytime sleep is associated with an increase of plasma PRL; acute reversal of the sleep activity cycle results in suppression of the nocturnal rise, but in production of a PRL daytime surge if subjects are allowed to sleep during daytime [3], suggesting that sleep is the main synchronizer of PRL circadian variations in man. The determinants of the small afternoon PRL surge are not elucidated.

Interrelations between sleep stages (or daytime basic rest activity cycles) and PRL secretory episodes remain controversial, since *Parker et al.* [4] reported a significant association between PRL episodic nadirs and rapid eye movements sleep periods, whereas other groups were not able to confirm such a relationship.

In the current study, plasma PRL profiles were obtained in 5 normal men submitted to westward and eastward transmeridian flights in order to evaluate adaptation of PRL variations to time shifts; sleep was monitored by polygraphic recordings throughout the study in order to detect possible correlations between PRL secretory episodes and sleep stages. Other pituitary and related hormones were measured simultaneously in each sample and will be reported elsewhere.

---

[1] This work was supported by the Belgian FRSM and APMO Foundation and was partially performed under contract of the Ministère Belge de la Politique Scientifique within the framework of the association Euratom-Universities of Brussels and Pisa.

Subjects and Methods

5 normal male subjects (aged 21–29 years) were submitted to seven studies (S1 to S7) at 10-day intervals during a total period of 10 weeks including a basal study (S1) in Brussels (Belgium), a westward 7-hour time shift by jet from Brussels to Chicago (USA), three studies (S2 to S4) performed one, 11 and 21 days after arrival in Chicago, an eastward flight back to Brussels and three studies (S5 to S7) one, 11 and 21 days after return in Brussels. During each study, blood samples were obtained at 15-min intervals for 25 h. During bedtime hours, blood was withdrawn through a long catheter extending to an adjoining room and sleep was monitored using standard polygraphic techniques. Immediately after arrival on each continent, the subjects were asked to comply strictly with the new local schedule for meals and sleep. Naps and recumbency during daytime were prevented.

Plasma PRL was measured in duplicate in each sample using a homologous human RIA, and was expressed in terms of the pituitary PRL research standard No. 71/222 of the MRC (UK): 1 μU of this standard is equivalent to 0.045 ng of the VLS standard.

The statistical analysis was performed using a method which has been recently developed by *Van Cauter* [5] for the characterization of blood components variations over the 24-hour span: periodogram calculations first provided a best-fit theoretical pattern, describing the low frequency components of the hormonal profile and accounting for its possible asymmetries; then acrophases and nadirs (respectively maximas and minimas of the best-fit pattern) were determined; finally, the profile was smoothed by eliminating all non-significant variations and the significant secretory peaks were defined as the peaks observed in the smoothed profile. Relation between plasma PRL and sleep stages was finally determined for each secretory peak using an approach similar to the method previously described by *Parker et al.* [4].

Results

No consistent changes in 24-hour mean levels of plasma PRL were detected from one study to another. Concentrations of PRL were significantly higher during bedtime than during wakefulness in all but one profile. The total number of peaks observed on the smoothed patterns ranged from 8 to 19, averaging 12 $\pm$ 3 per 24-hour period (mean $\pm$ SEM); the subjects spent 69 $\pm$ 3% of their time awake and exhibited during that time 71 $\pm$ 17% of their PRL secretory peaks; conversely, sleep accounted for 31 $\pm$ 6% of the whole investigation span and was associated with 29 $\pm$ 11% of secretory episodes. The frequency of PRL secretory episodes and their distribution between day and night were not affected by the transmeridian flights.

The 24-hour best-fit theoretical pattern of plasma PRL (fig. 1) was consistently found biphasic under basal conditions, with a minor acrophase peaking in the late afternoon and a more elevated acrophase occurring after the onset of sleep. Biphasic patterns were observed in each individual study

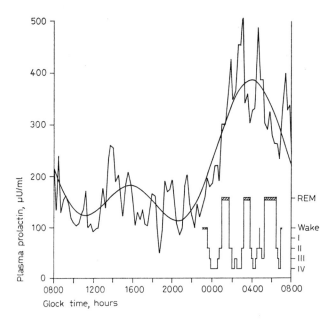

Fig. 1. Typical basal profile of plasma PRL in 1 individual, with saw-toothed variations of hormonal concentrations and superimposed biphasic best-fit theoretical pattern. Sleep stages are shown in the lower part.

under basal conditions and were largely predominant in studies performed at distance of flights.

Conversely, 6 out of 9 patterns obtained just after the flights exhibited splitting or disappearance of the afternoon acrophase and appeared either mono- or triphasic. 1 day after arrival on each continent, the major PRL acrophase already occurred at the expected time (i.e. during sleep). However, the adaptation of the acrophases was not fully achieved at that time: for the group of 5 subjects, mean sleep acrophase occurred significantly earlier in Chicago (S2) than in Brussels (S1) [01.45±00.58 vs 04.12±00.34 hours, local time, mean ± SD, p< 0.005]. A trend was observed for return to basal timing of wake and sleep acrophases (i.e. around 17.00 and 05.00 hours; local time, respectively) 11 days after arrival in Chicago, but full synchronization on basal patterns was only achieved in studies performed 21 days after arrival. After the eastward flight, the mean sleep acrophase was slightly delayed [03.48±00.56 vs 05.11±0.210 hours for S4 and S5, respectively, NS] but individual sleep acrophases showed a much larger time dispersion than in other studies.

Finally, the secretory episodes of PRL during sleep were confronted with polygraphic recordings. As expected, actual sleep onset was associated with marked elevation of plasma PRL in most profiles. In individual profiles, only 17% of secretory peaks occurred at the end of a REM period whereas 50% of the peaks occurred after well-characterized non-REM periods. Hormonal data of each individual profile were then synchronized from the end of each available REM period and transverse mean profiles were built up for each consecutive sleep cycle. No consistent relationship could be found in these transversal profiles between REM episodes and plasma PRL decrease or between onset of non-REM episodes and PRL release.

Comments

No quantitative alterations of mean 24-hour PRL concentrations were found throughout the study, suggesting that time shifts did not induce short-term stress effects on PRL secretion.

The nyctohemeral profiles of plasma PRL observed after transmeridian flights exhibited rough adaptation to the new local time already 1 day after arrival on each continent, confirming that prolactin periodicity is shifted immediately when an acute sleep/wake reversal occurs [2, 4]. However, small but persistent lag was observed for adaptation of the acrophases, with significant delay regarding expected time after the westward flight and slight advance after the eastward flight.

Theoretical best-fit patterns consistently disclosed a diurnal elevation of PRL plasma concentrations, with wakefulness acrophase occurring around 17.00 hours in basal studies. This afternoon acrophase accounted for the biphasic character of most patterns obtained at distance of flights. The splitting or disappearance of the wakefulness acrophase in patterns obtained immediately after the flights might indicate that the mechanisms governing diurnal and nocturnal increases of PRL secretion could be transiently dissociated after time shifts, and hence be controlled by different neuroendocrine modulators.

A steep elevation of plasma PRL was consistently observed after sleep onset, as previously reported by others [1, 2]. This predominant influence of sleep on PRL variations was not affected by time zone shifts. However, the inhibitory effect of REM sleep on PRL secretion described by *Parker et al.* [4] was not observed: PRL peaks were randomly distributed throughout sleep stages in basal studies as well as in studies performed after jet flights. This discrepancy with previous observations could be partially assigned to a lack of standardization in analysis of hormonal and sleep profiles: *Parker*'s data were based on plasma samples obtained at 20-min intervals and sleep recordings scored in 20-sec epochs, 60 of which must be summed up to score each interval between blood samples. In these conditions, sleep scoring is arbitrary and important variations should be expected from one laboratory to another. Nevertheless, *Parker et al.* [4] admitted that association between PRL nadirs and REM periods was of the 'usually but not always' type and unfortunately provided no figures to support this statement, but only transversal means of individual profiles showing this sleep stages-PRL peaks association. Even when our data, based on 15-min interval sampling, were similarly handled, they failed to reproduce any consistent effect of REM sleep upon individual PRL profiles.

References

1 Parker, D.C.; Rossman, L.G., and Vanderlaan, E.F.: Sleep-related, nyctohemeral and briefly episodic variations in human plasma prolactin concentrations. J. clin. Endocr. Metab. *36:* 1119–1124 (1973).
2 Sassin, J.F.; Frantz, A.G.; Kapen, S., and Weitzman, E.D.: The nocturnal rise of human prolactin is dependent on sleep. J. clin. Endocr. Metab. *37:* 436–440 (1973).
3 Frantz, A.G.: Rhythms in prolactin secretion; in Krieger, Endocrine rhythms, pp. 175–186 (Raven Press, New York 1979).
4 Parker, D.C.; Rossman, L.G., and Vanderlaan, E.F.: Relation of sleep-entrained human prolactin release to REM- and non-REM cycles. J. clin. Endocr. Metab. *38:* 646–651 (1974).
5 Van Cauter, E.: Method for characterization of 24-hour temporal variations of blood components. Am. J. Physiol. *237:* E255–E264 (1979).

D. Désir, MD, Service de Médecine Interne et Laboratoire de Médecine Expérimentale, Free University of Brussels, Hôpital Saint-Pierre, rue Haute 322 B-1000 Brussels (Belgium)

Discussion

Jacobs: I should simply like to confirm that, having also looked for the same REM/non-REM prolactin relationship, we were unable to confirm it. So, we agree completely with you. We looked very carefully in 8 subjects for 2 nights each. The only explanation that I can think of is that *Parker*'s data were very heavily weighted by a large number of nights from a small number of subjects in the overall group. There may have been undue influence of a certain number of subjects who may have shown that pattern.

Frantz: I have to say that we too were unable to find the same things that *Parker* found in our studies of 8 normal patients during sleep.

Regulation of Prolactin Secretion by Endogenous Sex Steroids in Man[1]

Maire T. Buckman, Laxmi S. Srivastava and Glenn T. Peake

Veterans Administration Hospital, University of New Mexico School of Medicine, Albuquerque, N.Mex., and University of Cincinnati School of Medicine, Cincinnati, Ohio

The role of endogenous sex steroids in the regulation of prolactin secretion in man remains unsettled. Few studies have evaluated the role of testosterone in the modulation of prolactin secretion [2, 26]. More extensive studies are available regarding the possible role of endogenous estrogen in the regulation of prolactin secretion, but conclusions from the various studies are frequently contradictory [3–8, 10–29]. Although the association between elevations in serum estrogen and prolactin concentrations during pregnancy has been established [1, 24], suggesting that there is a cause and effect relationship, the consequence of more modest changes or differences in estrogen levels on prolactin secretion have yielded diverse conclusions.

The current study was designed to further evaluate the possible role of endogenous estrogen and testosterone on perphenazine-induced prolactin responsiveness. All studies were performed on young adults. Three groups of studies were performed to evaluate estrogen effect: (1) comparison of prolactin responsiveness during short-term spontaneous fluctuations in estrogen during normal menstrual cycles; (2) comparison of prolactin responsiveness during chronically low vs intermittently low estrogen states; and (3) evaluation of the effect of very low serum estrogen levels on prolactin responsiveness. To study the possible modulatory role of testosterone on prolactin secretion, the effect of high vs low endogenous testosterone concentrations on prolactin responsiveness was examined.

---

[1] This research was supported by the Veterans Administration Clinical Investigator Award to Dr. *Buckman* and by NIH Grants HD 05794–08 and RR 00997-04.

Materials and Methods

Subjects

All subjects volunteered to participate in the studies by informed consent. None had been on chronic medication for 3 months prior to the studies. Four groups were recruited:

(1) 12 normal women with $\bar{x} \pm 1$ SE age of 26 ± 2 years without overt or occult galactorrhea and with regular menses (cycle interval 25–32 days) agreed to participate in the studies [7]. The women were studied between days 2–6 (early follicular phase = EFP) and again between days 11–16 (late follicular phase = LFP) of the menstrual cycle. The LFP period was confirmed by serum progesterone measurement which was in the follicular phase range for each subject.

(2) 13 women ($\bar{x}$ age 24 ± 2 years) with secondary amenorrhea of greater than 6 months duration were included in the study. All women were free of systemic illness. Physical examination was normal in each woman; none had hirsutism or acne. Thyroid function studies and basal prolactin were normal in each woman. Serum LH and FSH were normal or low in each, excluding the possibility of ovarian insufficiency. To exclude the presence of a sellar or parasellar mass lesion, careful sella polytomography at 1-mm intervals in the sagittal and coronal planes and visual fields by Goldman perimetry were performed. None of the patients exhibited abnormalities of the sella turcica or visual field defects. Onset of amenorrhea, as far as could be determined, was associated with withdrawal of oral contraceptives, simple weight loss or psychological stress. Following completion of the studies, 3 of the women have spontaneously resumed regular menstrual cycles, suggesting the presence of a reversible hypothalamic-pituitary dysfunctional state.

(3) 17 normal men ($\bar{x}$ age 24 ± 1 years) participated in the studies.

(4) 7 boys ($\bar{x}$ age 18 ± 1 years) with idiopathic delayed puberty were studied. None had evidence of systemic illness. Physical examination was unremarkable in each except for sexual immaturity. Buccal smear revealed a normal male pattern in each individual excluding Klinefelter's syndrome. Thyroid function studies were normal in each boy. Serum LH, FSH and testosterone were low in each subject confirming the diagnosis of hypogonadotropic hypogonadism. Sella polytomography and visual field examination by Goldman perimetry were normal in each patient, excluding a significant sellar or parasellar mass lesion.

Perphenazine Test

All studies were performed after an overnight fast. 8 mg perphenazine (Trilafon®) was administered orally between 0800 and 1000 hours. Blood samples were obtained via intravenous cannula or by venipuncture twice before and 1, 2, 3, 4, 5, and 6 h following perphenazine ingestion. Subjects were allowed normal activity but no vigorous exercise for the duration of the study.

Steroid and Prolactin Assays

Steroid levels were measured on serum obtained at time '0' of the perphenazine test. Estrone (E_1), estradiol (E_2), and progesterone (p) were determined as previously described [7]. Serum testosterone was determined according to the method described by *Coyotupa et al.* [9]. Utilizing this method for testosterone assay, the inter- and intrassay coefficients of variation were 9.4 and 4.8%, respectively.

Table I. Serum steroid levels ($\bar{x} \pm 1$ SE)

Group	(n)	Age years	E_1 pg/ml	E_2 pg/ml	P pg/ml	T ng/ml	PRL response ng·h/ml
Normal women, EFP	(12)	26 ± 2	42 ± 3	40 ± 6	217 ± 26	–	317 ± 46
Normal women, LFP	(12)	26 ± 2	118 ± 18a	203 ± 49a	198 ± 29	–	402 ± 46b
Amenorrheic women	(13)	24 ± 2	49 ± 4	31 ± 3	220 ± 33	0.5 ± 0.1	142 ± 18c
Hypogonadal men	(7)	18 ± 1	24 ± 2d	19 ± 1d	311 ± 53	0.6 ± 0.1	149 ± 26
Normal men	(17)	24 ± 1	47 ± 6	36 ± 3	246 ± 23	8.1 ± 0.8d	122 ± 12

a $p < 0.005$ vs EFP; b $p < 0.02$ vs EFP; c $p < 0.005$ vs normal women, EFP; d $p < 0.001$ vs amenorrheic women.

Prolactin was assayed as previously described [7]. All samples from a single individual were run in duplicate in the same assay. To minimize interassay variation, samples from groups of subjects whose responses were being compared were run in the same assay.

Statistical Analyses
For each perphenazine test, integrated prolactin response was determined over the 6-hour study period using a program for trapezoidal integration. The two-tailed paired or unpaired Student's t-test was used for statistical analysis.

Results

Serum progesterone concentrations were similar in all 5 study groups (table I). Thus, the possible regulatory role of this steroid on prolactin secretion was not evaluated in the present study.

Effect of Estrogen (table I, fig. 1)
(1) *Comparison of Prolactin Responsiveness during Short-Term Spontaneous Fluctuations in Estrogen during Normal Menstrual Cycles.* Model: Normal women during the low (EFP) and high (LFP) estrogen phases of the menstrual cycle [7]. Serum E_1 and E_2 were significantly higher during the LFP compared to the EFP of the cycle. Progesterone levels were comparable in the 2 groups confirming that the studies were performed in the pre-

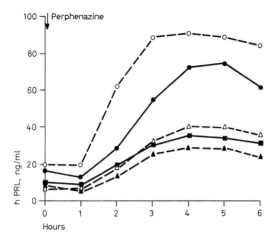

Fig. 1. Mean prolactin concentrations before and at hourly intervals following perphenazine administration in 5 groups of subjects. '0' time represents the mean of 2 baseline levels. ○ = Normal women during the LFP; ● = normal women during the EFP; △ = amenorrheic women; ■ = hypogonadal men; ▲ = normal men. Sex steroid concentrations for each group are shown in table I.

ovulatory phase of the cycle. The increased concentration of estrogen during the LFP was associated with a significantly enhanced perphenazine-induced prolactin response. The mean change in integrated prolactin response was +37% with a range of −24 to +81%. Thus, the higher estrogen state was associated with facilitated perphenazine-induced prolactin secretion in the majority of subjects studied, suggesting that estrogen may facilitate prolactin secretion in normal women during the LFP period.

(2) Comparison of Prolactin Responsiveness during Chronically Low vs Intermittently Low Estrogen States. Model: Amenorrheic women vs normal women during the EFP of the cycle. Serum estrogen levels were similar in these 2 study groups. The integrated prolactin response, however, was significantly greater in the normal women during the EFP than in the chronically amenorrheic women. This observation suggests that the long-term integrated estrogen milieu may play an important role in determining the magnitude of perphenazine-induced prolactin responsiveness. Cyclic hyperestrogenism in normally cycling women continued to prime the pituitary mammotropes even during a period when measured serum estrogen levels were low and in the same range as those measured in the amenorrheic group.

(3) *Evaluation of the Effect of Very Low Serum Estrogen Levels on Prolactin Responsiveness.* Model: Hypogonadal males vs amenorrheic women. Serum E_1 and E_2 levels were significantly lower in the hypogonadal males compared to the chronically amenorrheic women. Perphenazine-induced prolactin responsiveness, however, was similar in the 2 groups. This observation suggests that below a critical estrogen milieu, estrogens may not play an important role in the regulation of perphenazine-induced prolactin secretion. Other regulatory factors, as yet undefined, may maintain prolactin responsiveness in the face of very low estrogen levels. Serum testosterone was measured in these 2 groups of individuals and was found to be comparable in the 2 groups.

Effect of Testosterone (table I, fig. 1)

Model: Normal men vs amenorrheic women. To evaluate the possible role of testosterone in the regulation of prolactin secretion, perphenazine-induced prolactin responsiveness was compared in a group of individuals with chronically low serum testosterone levels (amenorrheic women) and in a group of subjects with chronically high serum testosterone concentration (normal men). Serum E_1 and E_2 were similar in the 2 groups thus excluding any possible modulatory effect derived from differences in estrogen levels. The mean integrated prolactin response was lower in the normal men but this difference was not statistically significant. Thus, if high concentrations of testosterone exert an inhibitory effect on prolactin secretion, the effect is modest in amplitude and in our study was not found to be statistically significant.

Discussion

Contradictory data exist regarding the role of endogenous estrogen in the regulation of prolactin secretion in man. One of the experimental approaches to this problem has been the evaluation of single basal and stimulated prolactin secretion during cyclic changes in endogenous estrogen associated with the normal menstrual cycle. Daily measurement of basal prolactin levels throughout menstrual cycles in normal women have yielded data which suggest that estrogen may not influence prolactin levels [14, 17, 19, 28] or that there is a parallel change in serum estrogen and prolactin concentrations [12, 29]. *Franchimont et al.* [12] further demonstrated that there was a significant correlation between the percent increase in serum estradiol and prolactin during ovulatory and luteal compared to follicular phases of the cycle. The reasons for the discrepancies observed in the various studies measuring prolactin on single random blood specimens may be multifactorial. Factors other than estrogen may affect random prolactin levels.

Prolactin concentrations may be influenced by changes due to diurnal variation, spontaneous fluctuations in prolactin secretion and stress-induced elevations in serum prolactin concentration [8]. Since changes in prolactin, when observed, are relatively modest throughout the menstrual cycle: 32% [12] to 50% [29] increase in periovulatory vs EFP levels and 19% in luteal vs EFP levels [12], other factors, independent of changes in serum estrogen concentrations, may obscure these changes.

To control for factors known to influence prolactin concentrations such as diurnal variation, spontaneous fluctuations in prolactin secretion, and stress which may obscure a modest estrogen effect, we measured serum prolactin levels every 15 min for 5 h following a 30-min equilibration period during the early and again during the periovulatory phase of 10 normal menstrual cycles [8]. Mean prolactin concentrations for individual subjects were consistently higher during the periovulatory compared to the early follicular phase study period. The mean prolactin of the means for the individual subjects increased from 8.1 ± 1.0 during the low estrogen phase to 10.7 ± 1.0 ng/ml during the high estrogen phase of the cycle ($p < 0.001$). The mean increase in prolactin for the group was 39% with a range of 9–117%. Similar to the observation previously published by *Franchimont et al.* [12], the incremental change in prolactin over the 2 study periods correlated with the incremental change in estradiol but not with the incremental change in estrone, progesterone or 17-hydroxyprogesterone. Thus, this study suggests that enhancement of endogenous ovarian estrogen secretion during the periovulatory phase was associated with a significant but variable enhancement of spontaneous prolactin secretion which correlated with the incremental change in estradiol.

In addition to the determination of basal and spontaneous prolactin patterns throughout normal menstrual cycles, stimulated prolactin responses have been evaluated during various endogenous estrogen states of the cycle. TRH-induced prolactin responses have been reported to be similar during the EFP, periovulatory and luteal phases of the menstrual cycle [20, 25, 28]. In contrast, TRH-induced prolactin responses have been reported to be significantly enhanced during the periovulatory [23] and luteal [5] compared to the EFP of the menstrual cycle. As with the previously discussed studies, the differences observed during the higher endogenous estrogen states were modest in amplitude.

Buckman et al. [7] describe here perphenazine-induced prolactin responses in normal women during EFP and LFP of the menstrual cycle. Prolactin responsiveness was variably enhanced in 11 of 12 studies with a mean increase in integrated prolactin secretion of 37% during the higher endogenous estrogen phase. Thus, basal and pharmacologically stimulated prolactin concentrations in those studies in which enhancement was observed was increased 50% or less during the higher compared to the lower endogenous estrogen phase.

In addition to evaluating prolactin concentrations during short-term changes in ovarian estrogen secretion throughout normal menstrual cycles, random prolactin concentrations have been determined in various populations of individuals and correlated with the ambient estrogen milieu. Basal prolactin concentrations were found to be similar in children, adult men and adult women [13]. Most other studies, however, have reported a small but significant elevation in mean serum prolactin in women compared to men [15, 21]. Furthermore, males with relatively more stable estrogen concentrations than women have been shown to have stable prolactin levels throughout their life span [24]. In girls, puberty has been associated with parallel changes in serum estrogen and prolactin levels [27]; however, an additional study failed to confirm this relationship [18]. Postmenopausal women have been reported to have lower prolactin levels than women during the reproductive period of their lives [24]. Thus, correlation of endogenous estrogen exposure with prolactin levels has, for the most part, suggested that estrogens may facilitate prolactin secretion, but as with the previously discussed studies, negative reports also exist. Perhaps similar pitfalls contribute to the difficulty in interpretation of single random blood prolactin levels as were discussed in reference to prolactin concentrations throughout normal menstrual cycles.

Prolactin responsiveness following administration of prolactin secretagogues has been reported in various populations of individuals. TRH-induced prolactin responses have been found to be enhanced in women compared to men [4, 22]. Likewise, phenothiazine-induced responses are greater in women than in men [6]. In addition, postmenopausal women have diminished prolactin responses to a variety of prolactin secretagogues compared to women in the reproductive age group [3, 11, 16]. These data suggest that the differences in ambient estrogen milieus may be responsible for the differential prolactin responses observed.

In the current study the effect of various estrogen milieus on prolactin responsiveness was evaluated following administration of the phenothiazine, perphenazine. To correct for a possible age-associated variability in prolactin reserve, young adults were chosen for all studies. Amenorrheic women with chronically stable, low estrogen concentrations were compared to normal women with cyclic estrogen secretion during the EFP of the cycle when measured estrogen levels were similar to those measured in the amenorrheic group. Perphenazine-induced prolactin responsiveness was significantly enhanced in the women with cyclic estrogen secretion. This observation suggests that cyclic increase in estrogen secretion may have a long-term effect on enhancement of prolactin secretion even during states of relatively low circulating estrogen levels. The second series of studies evaluated the effect of very low chronic estrogen levels on prolactin secretion. Although measured estradiol levels were 40% lower in hypogonadal men compared to amenorrheic women, the perphenazine-induced prolactin responses were similar in the 2 groups. This observation suggests that below a certain critical serum estrogen level, estrogens may not play an important role in the maintenance of phenothiazine-induced prolactin secretion.

Little is known of the role of testosterone in the regulation of prolactin secretion in man. Orchiectomy has been reported not to influence prolactin secretion in men with prostatic carcinoma [2] suggesting that testosterone may not modulate serum prolactin levels. In a recent study by *Spitz et al.* [26] an inverse correlation between basal testosterone and peak TRH-induced prolactin concentrations was observed. This study implies that testosterone may suppress TRH-induced prolactin responses.

The current study examined the perphenazine-induced prolactin response in 2 groups of subjects with similar serum estrogen levels but vastly different testosterone concentrations. Although the mean prolactin response was modestly blunted in normal men compared to amenorrheic women, the difference failed to achieve statistical significance. We conclude that serum testosterone may not play a major role in the regulation of prolactin secretion in man.

Acknowledgements

We wish to acknowledge the assistance of the staff of the University of New Mexico Clinical Research Center in performing the studies, the technical expertise of *Barry David* and *Bharat Shah*, and the expert and patient secretarial assistance of *Maxine Payne*.

We thank Drs. *VanderLaan, Lewis* and *Sinha* and the National Pituitary Agency, NIAMDD, for giving us the materials for the homologous human prolactin radioimmunoassay.

Summary

The current study sought to determine the role of the endogenous sex steroids, estrogen and testosterone, on phenothiazine-induced prolactin responsiveness in man. We conclude from these studies that (1) Cyclic ovarian estrogen secretion in normal women results in an enhancement of spontaneous and phenothiazine-induced prolactin release during the high compared to the low estrogen phase of the cycle. (2) Cyclic estrogen secretion enhances prolactin release even during periods of relatively low serum estrogen concentrations suggesting that the long-term estrogen milieu has a major influence on mammotrope responsiveness. (3) Below a critical estrogen milieu, equivalent to that seen in normal adult men, factors other than estrogen play a role in the maintenance of prolactin secretion and (4) Endogenous testosterone may not play a major role in modulating prolactin secretion in man.

References

1 Barberia, J.M.; Abu-Fadil, S.; Kletzky, O.A.; Nakamura, R.M., and Mishell, D.R.: Serum prolactin patterns in early human gestation. Am. J. Obstet. Gynec. *121:* 1107–1110 (1975).
2 Bartsch, W.; Horst, H.J., and Nehse, G.: Sex hormone binding globulin binding capacity, testosterone, 5-alpha-dihydrotestosterone, oestradiol and prolactin in plasma of patients with prostatic carcinoma under various types of hormonal treatment. Acta endocr., Copenh. *85:* 650–664 (1977).
3 Bohnet, H.G.; Greiwe, M.; Hanker, J.P.; Aragona, C., and Schneider, H.P.G.: Effects of cimetidine on prolactin, LH, and sex steroid secretion in male and female volunteers. Acta endocr., Copenh. *88:* 428–434 (1978).
4 Bowers, C.Y.; Friesen, H.G.; Hwang, P.; Guyda, H.J., and Folkers, K.: Prolactin and thyrotropin release in man by synthetic pyroglutamyl-histidyl-prolinamide. Biochem. biophys. Res. Commun. *45:* 1033–1041 (1971).
5 Boyd, A., III and Sanchez-Franco, F.: Changes in the prolactin response to thyrotropin-releasing hormone (TRH) during the menstrual cycle of normal women. J. clin. Endocr. Metab. *44:* 985–989 (1977).
6 Buckman, M.T. and Peake, G.T.: Dynamic evaluation of prolactin secretion with perphenazine in normal and hyperprolactinemic subjects. Hormone metabol. Res. *10:* 365–458 (1978).
7 Buckman, M.T.; Peake, G.T., and Srivastava, L.S.: Endogenous estrogen modulates phenothiazine stimulated prolactin secretion. J. clin. Endocr. Metab. *48:* 901–906 (1976).

8 Buckman, M.T.; Peake, G.T., and Srivastava, L.S.: Periovulatory enhancement of spontaneous prolactin secretion in normal women. Metabolism 29: 753–757 (1980).
9 Coyotupa, J.; Parlow, A.F., and Abraham, G.E.: Simultaneous radioimmunoassay of plasma testosterone and dihydrotestosterone. Analyt. Lett. 5: 329–340 (1972).
10 Ehara, Y.; Siler, T.; Vandenberg, G.; Sinha, Y., and Yen, S.: Circulating prolactin levels during the menstrual cycle: episodic release and diurnal variation. Am. J. Obstet. Gynec. 117: 962–970 (1973).
11 Franchimont, P.; Dourcy, G.; Legros, J.J.; Reuter, A.; Vrindts-Gevaert, Y.; Cauwenberge, J.R. van; Remacle, P.; Gaspard, U. et Colin, C.: Dosage de la prolactine dans les conditions normales et pathologiques. Annls Endocr. 37: 127 (1976).
12 Franchimont, P.; Dourcy, C.; Legros, J.; Reuter, A.; Vrindts-Gevaert, Y.; Cauwenberge, J., and Gaspard, U.: Prolactin levels during the menstrual cycle. Clin. Endocrinol. 5: 643–650 (1976).
13 Friesen, H.G. and Hwang, P.: Human prolactin. A. Rev. Med. 24: 251–270 (1973).
14 Hwang, P.; Guyda, H., and Friesen, H.: A radioimmunoassay for human prolactin. Proc. natn. Acad. Sci. USA 68: 1902–1906 (1971).
15 Jacobs, L.S.; Mariz, I.K., and Daughaday, W.H.: A mixed heterologous radioimmunoassay for human prolactin. J. clin. Endocr. Metab. 33: 1–7 (1971).
16 Jacobs, L.S.; Snyder, P.S.; Utiger, R.D., and Daughaday, W.H.: Prolactin response to thyrotropin-releasing hormone in normal subjects. J. clin. Endocr. Metab. 36: 1069–1073 (1973).
17 Jaffe, R.; Yuen, B.; Keye, W., and Midgley, A.: Physiologic and pathologic profiles in circulating human prolactin. Am. J. Obstet. Gynec. 117: 757–773 (1973).
18 Lee, P.A.; Xenakis, T.; Winer, J., and Matsenbaugh, S.: Puberty in girls: correlation of serum levels of gonadotropins, prolactin, androgens, estrogens, and progestins with physical changes. J. clin. Endocr. Metab. 43: 775–783 (1976).
19 McNeilly, A.; Evans, G., and Chard, T.: Observations of prolactin levels during the menstrual cycle; in Pasteels and Robyn, Human prolactin, pp. 231–232 (Excerpta Medica, Amsterdam 1973).
20 McNeilly, A. and Hagen, C.: Prolactin, TSH, LH and FSH responses to a combined LHRH/TRH test at different stages of the menstrual cycle. Clin. Endocrinol. 3: 427 (1974).
21 Nokin, J.; Vekemans, M.; L'Hermite, M., and Robyn, C.: Circadian periodicity of serum prolactin concentration in man. Br. med. J. iii: 561–562 (1972).
22 Rakoff, J.S.; Siler, T.M.; Sinha, Y.N., and Yen, S.S.C.: Prolactin and growth hormone release in response to sequential stimulation by arginine and synthetic TRH. J. clin. Endocr. Metab. 37: 641–644 (1973).
23 Reymond, M. and Lemarchand-Béraud, T.: Effects of oestrogens on prolactin and thyrotropin responses to TRH in women during the menstrual cycle and under oral contraceptive treatment. Clin. Endocrinol. 5: 429–437 (1976).
24 Robyn, C.; Delvoye, P.; Exter, C. van; Vekemans, M.; Caufriez, A.; Denayer, P.; Delogne-Desnoeck, J., and L'Hermite, M.: Physiological and pharmacological factors influencing prolactin secretion and their relation to human reproduction; in Crosignani and Robyn, Prolactin and human reproduction, pp. 71–96 (Academic Press, London 1977).
25 Sawin, C.T.; Hershman, J.M.; Boyd, A.E., III; Longcope, C., and Bacharach, P.: The relationship of changes in serum estradiol and progesterone during the menstrual

cycle to the thyrotropin and prolactin responses to thyrotropin-releasing hormone. J. clin. Endocr. Metab. *47:* 1296–1302 (1978).

26 Spitz, I.M.; Zylber, E.; Cohen, H.; Almaliach, U., and Leroith, D.: Impaired prolactin response to thyrotropin-releasing hormone in isolated gonadotropin deficiency and exaggerated response in primary testicular failure. J. clin. Endocr. Metab. *48:* 941–945 (1979).

27 Thorner, M.O.; Round, J.; Fahmy, D.; Groom, G.V.; Butcher, S., and Thompson, K.: Serum prolactin and oestradiol levels at different stages of puberty. Clin. Endocrinol. *7:* 463–468 (1977).

28 Tyson, J. and Friesen, H.: Factors influencing the secretion of human prolactin and growth hormone in menstrual and gestational women. Am. J. Obstet. Gynec. *116:* 377–387 (1973).

29 Vekemans, P.; Delvoye, M.; L'Hermite, M., and Robyn, C.: Serum prolactin levels during the menstrual cycle. J. clin. Endocr. Metab. *44:* 989–993 (1977).

M.T. Buckman, MD, Research Service (151), Veterans Administration Hospital, 2100 Ridgecrest Drive S.E., Albuquerque, NM 87108 (USA)

Control of the Hypoglycemia Release of Prolactin

M.D. Whitaker, B. Corenblum, P.J. Taylor and P.H. Harasym

Health Sciences Centre, The University of Calgary, Departments of Medicine, Obstetrics and Gynecology, and Educational Assessment and Planning, Calgary, Alberta

In humans, it appears that prolactin secretion is under the predominant influence of a hypothalamic prolactin-inhibiting factor (PIF), believed to be dopaminergic in nature. Several studies suggest the existence of a separate prolactin-releasing factor (PRF), but presently there is little agreement on the nature of this substance. The most widely accepted view is that the PRF is released from the brain with subsequent action on the pituitary. Various neurotransmitters have been studied, more in rodents than in primates, and the nature of the neurotransmitter control of PRF is equally controversial. One of the most widely studied neurotransmitters has been serotonin but there is a paucity of such data in humans. Nevertheless, serotonin antagonists, such as cyproheptadine [16], have been used with some success in clinical reproductive disorders associated with altered prolactin secretion. There are several situations in humans where a sudden increase in circulating serum prolactin may be induced, and one reproducible stimulus is insulin-induced hypoglycemia. The purposes of these studies were to determine whether the release of prolactin by hypoglycemia is mediated by inhibition of the dopaminergic PIF, or rather mediated by release of a separate PRF which is not mediated via the dopaminergic pathway. Furthermore, if evidence suggests a PRF, then to further define the controlling neurotransmitter.

Methods

Prolactin release was produced by insulin-induced hypoglycemia, performed 2 h after awakening. Insulin, 0.12 U/kg, was injected i.v. into an indwelling venous catheter. Serum prolactin and glucose levels were measured at −30, 0, 30, 45, 60, and 90 min. If the test was performed while the volunteer was taking a medication, then the last tablet was taken 2 h prior to the insulin tolerance test. Serum glucose fell to less than 35 mg/dl in all tests. At 90 min post-insulin injection, TRH 200 µg was administered as an i.v. bolus, and serum prolactin was drawn 20 min later.

Four studies were performed. In the first study cyproheptadine was studied in 6 normoprolactinemic women who had amenorrhea 6 months after discontinuation of oral contraceptive therapy. After a control insulin tolerance test (ITT) was performed, cyproheptadine 8 mg every 7 h for 21 h was administered and the ITT repeated.

In the second study, 8 normal volunteers (6 females studied in the mid-follicular phase of the menstrual cycle, and 2 males) had a control ITT-TRH test, and were then administered pimozide 2 mg twice a day by mouth for 48 h, and then the ITT-TRH test repeated. Cyproheptadine, 8 mg every 7 h, was then added to the pimozide regimen and both drugs were given for a further 24 h at which time the ITT-TRH test was repeated.

The third study was performed in 4 normal female volunteers studied in the mid-follicular phase of the menstrual cycle. A control ITT test was performed, and then chlorpheniramine maleate 8 mg was administered orally every 7 h for 21 h, and then the ITT repeated.

The fourth study was performed in 4 normal females studied in the mid-follicular phase. After a control ITT-TRH test pimozide, 2 mg twice daily, and chlorpheniramine, 8 mg every 7 h, were both administered over 21 h, and the ITT-TRH test repeated.

Serum prolactin was measured by a double antibody radioimmunoassay. The increment of response of serum prolactin with regard to each baseline prolactin concentration was statistically analyzed by a two-way analysis of variance with repeated measures on two factors, treatment and time. In addition, a test of simple mean effects was carried out.

Results and Discussion

Cyproheptadine was chosen as it is a recognized serotonin antagonist. It has been shown to block the release of prolactin produced by drugs directly stimulating the serotonin receptor [10]. In man it has been shown to antagonize the increase of serotonin produced by oral administration of 5-hydroxytryptophan [9]. In the first study, as in all other studies, peak prolactin response was seen at 60 min post-insulin injection, and this peak level minus the basal level was defined as the increment of response. Cyproheptadine did not alter the basal prolactin level but did significantly blunt the increment of prolactin release in response to hypoglycemia (table I).

To further examine this inhibition of release it was necessary to assess whether this involved dopaminergic blockade (i.e. inhibition of the PIF) and for this reason the dopaminergic pathway had to be inhibited. In the second study, dopaminergic blockade, achieved with pimozide, did significantly increase the basal prolactin level, and the increment of response to hypoglycemia was also significantly blunted. The further addition of cyproheptа-

Table I. Basal and maximal increment response of serum prolactin (ng/ml) to hypoglycemia before and after cyproheptadine (mean ± SE)

	Basal	Increment	
Control	5.2 ± 0.9	58.3 ± 12.6	(1)
Cyproheptadine	5.7 ± 0.8	27.0 ± 6.0	(2)

(1) vs (2) $p < 0.01$.

Table II. Response of serum prolactin (ng/ml) to hypoglycemia with control, pimozide, and pimozide plus cyproheptadine administration; and further response to TRH (mean ± SE)

	Basal		Increment		TRH
Control	5.8 ± 0.7	(1)	49.4 ± 7.0	(2)	20 ± 7
Pimozide	26.3 ± 3.3	(3)	24.1 ± 5.9	(4)	19 ± 6
Pimozide plus cyproheptadine	30.6 ± 5.8	(5)	12.0 ± 5.1	(6)	32 ± 10

(3) vs (5) n.s., (2) vs (4) $p < 0.05$., (4) vs (6) $p < 0.05$.

dine to the dopaminergic blockade did not change the basal prolactin level, but the increment of response to hypoglycemia was significantly decreased compared to that with pimozide alone (table II). The release of prolactin in response to TRH occurred in all three study situations (table II), implying that the remaining pituitary prolactin reserve on all test days was not depleted by the dopaminergic blockade and therefore this was not the reason for the decreased incremental response to the hypoglycemic stimulus. These data therefore suggest that, under conditions of insulin-induced hypoglycemia, inhibition of a dopaminergic pathway does not completely account for the acute release of prolactin and another factor is involved.

The inhibition of prolactin secretion in response to hypoglycemia by cyproheptadine is potentially multi-factorial. Possible mechanisms of action of cyproheptadine are anti-serotoninergic, anti-histaminic, anti-cholinergic, anti-dopaminergic, or by an action exerted directly on the lactotroph. It is

Table III. Basal and maximal increment response of serum prolactin (ng/ml) to hypoglycemia before and after chlorpheniramine

	Basal	Increment	
Control	5.3 ± 0.8	45.4 ± 5.6	(1)
Chlorpheniramine	6.2 ± 1.1	9.4 ± 2.9	(2)

(1) vs (2) $p < 0.01$.

unlikely that the anti-cholinergic action of cyproheptadine is accounting for these results since data obtained in animals suggest that cholinergic drugs inhibit prolactin secretion probably by activation of hypothalamic catecholaminergic factors [8]. There is little human work but atropine failed to alter basal or insulin-stimulated prolactin levels in humans [2]. The anti-dopaminergic action of cyproheptadine is quite weak, and in monkeys was only found to acutely increase prolactin release with large intravenous dosages, and this effect was not seen with the oral route of administration of cyproheptadine [4]. Furthermore, the clinical utilization of cyproheptadine to treat women with hyperprolactinemic amenorrhea with resultant normalization of serum prolactin and resumption of menstrual activity [16] also suggests that the anti-dopaminergic action of cyproheptadine does not account for our results.

Some studies have found an action of cyproheptadine to be exerted directly at the level of the pituitary with resultant inhibition of prolactin release [11]. If this *in vitro* mechanism of action was accounting for the inhibition of prolactin release in our studies, then this action could only be present during the augmented release of prolactin, since the basal level of prolactin in both studies was not altered by cyproheptadine administration. Nevertheless, this potential mechanism of action must be considered and cannot be discounted at this time.

The anti-histaminic action of prolactin must be considered since histamine has been noted in many studies to acutely release prolactin, and this effect has been blocked by H_1 antagonists [1]. The effect of chlorpheniramine, a relatively pure H_1 antagonist, is shown in table III. Chlorpheniramine did not change the basal prolactin level but did significantly blunt the response to hypoglycemia. The addition of pimozide to the chlorpheniramine (table IV) did increase the basal serum prolactin level, as expected. The blunting of

Table IV. Basal and maximal increment response of serum prolactin (ng/ml) to hypoglycemia and to subsequent TRH before and after chlorpheniramine plus pimozide

	Basal	Increment	TRH
Control	7.4 ± 1.1	40.2 ± 8.9 (1)	31 ± 16
Chlorpheniramine plus pimozide	25.1 ± 3.9	32.4 ± 7.8 (2)	53 ± 18

(1) vs (2) p = n.s.

the increment of prolactin in response to hypoglycemia, previously seen with chlorpheniramine alone, was no longer present with the addition of pimozide to the chlorpheniramine. It appears that the inhibition of prolactin release by chlorpheniramine may require an intact dopaminergic pathway, and thus is a quite different response than that seen with cyproheptadine. With cyproheptadine the inhibition of prolactin release by hypoglycemia was not altered by dopaminergic blockade. These results may not be surprising as previous reports have found that the histamine-induced increase of prolactin may involve the dopaminergic pathway [6] and has been found in other studies to be reversed with adrenergic blockade [15].

We believe that the most likely mechanism of action by cyproheptadine accounting for our results is via an anti-serotoninergic mechanism. Recent studies support this conclusion. Insulin has been found to increase CNS serotonin content [7] and the hypoglycemia release of prolactin is greater in the evening than in the morning, in keeping with the known diurnal variation in hypothalamic serotonin activity [14]. Fluoxetine, a specific inhibitor of serotonin re-uptake (which would result in enhanced serotoninergic activity) was found not to effect the basal prolactin level but did significantly enhance the insulin-induced release of prolactin [13].

Because serotonin has been found not to have any direct action on the pituitary, we believe these data are most in keeping with concepts supported by other studies in that serotonin acts as a neurotransmitter to release a PRF from the brain which then acts on the pituitary to release prolactin [3, 5]. The serotonin release of prolactin has not been found to involve PIF activity [12] and the serotoninergic release of a PRF is not influenced by blockade of the catecholaminergic system [3]. Nevertheless, caution should be exercised in interpreting our results in view of the effect of cyproheptadine on several neurotransmitters as well as an effect directly on the lactotroph.

References

1 Arakelian, M.D. and Libertun, C.: H_1 and H_2 histamine receptor participation in the brain control of prolactin secretion in lactating rats. Endocrinology *100:* 890–895 (1977).
2 Boyd, A.E. and Reichlin, S.: Neural control of prolactin secretion in man. Psychoneuroendocrinology *3:* 113–130 (1978).
3 Clemens, J.A.; Roush, M.E., and Fuller, R.W.: Evidence that serotonin neurons stimulate secretion of prolactin releasing factor. Life Sci. *22:* 2209–2214 (1978).
4 Gala, R.R.; Peters, J.A.; Pieper, D.R., and Campbell, M.D.: Influence of adrenergic antagonists and apomorphine on prolactin release induced by serotoninergic antagonists in the monkey. Life Sci. *22:* 25–30 (1978).
5 Garthwaite, T.L. and Hagen, T.C.: Evidence that serotonin stimulates a prolactin releasing factor in the rat. Neuroendocrinology *29:* 215–220 (1979).
6 Gibbs, D.M.; Plotsky, P.M.; Greef, W.J. de, and Neill, J.D.: Effect of histamine and acetylcholine on hypophysial stalk plasma dopamine and peripheral plasma prolactin levels. Life Sci. *24:* 2063–2070 (1979).
7 Gordon, A.E. and Meldrum, B.S.: Effect of insulin on brain 5-hydroxytryptamine and 5-hydroxy indole acetic acid of rat. Biochem. Pharmacol. *19:* 3042–3044 (1970).
8 Grandison, L. and Meites, J.: Evidence of adrenergic mediation of cholinergic inhibition of prolactin release. Endocrinology *99:* 775–779 (1976).
9 Kato, Y.; Nakai, Y., and Imura, H.: Effect of 5-hydroxytryptophan (5-HTP) on plasma prolactin levels in man. J. clin. Endocr. Metab. *28:* 695–697 (1974).
10 Krulich, L.; Vijayan, E.; Coppings, R.J.; Giachetti, A.; McCann, S.M., and Mayfield, M.A.: On the role of the central serotoninergic system in the regulation of the secretion of thyrotropin and prolactin: thyrotropin-inhibiting and prolactin-releasing effects of 5-hydroxytryptamine and quipazine in the male rat. Endocrinology *105:* 276–283 (1979).
11 Lamberts, S.W.J. and MacLeod, R.M.: The effects of cyproheptadine on prolactin secretion by whole pituitary glands and dispersed pituitary tumor cells. Endocrinology *103:* 1710–1717 (1978).
12 Lu, K.H. and Meites, J.: Effects of serotonin precursors and melatonin on serum prolactin release in rats. Endocrinology *93:* 152–155 (1973).
13 Masala, A.; Delitala, G.; Devilla, L.; Alagna, S., and Rovasio, P.P.: Enhancement of insulin-induced prolactin secretion of fluoxetine in man. J. clin. Endocr. Metab. *49:* 350–352 (1979).
14 Nathan, R.S.; Sachar, E.J.; Langer, G.; Tabrizi, M.A., and Halpern, F.S.: Diurnal variation in the response of plasma prolactin, cortisol, and growth hormone to insulin-induced hypoglycemia in normal men. J. clin. Endocr. Metab. *49:* 231–235 (1979).
15 Rivier, C. and Vale, W.: Effects of γ-aminobutyric acid and histamine on prolactin. Endocrinology *101:* 506–511 (1977).
16 Wortsman, J.; Soler, N.G., and Hirschowitz, J.: Cyproheptadine in the management of the galactorrhea-amenorrhea syndrome. Ann. intern. Med. *90:* 923–925 (1979).

M.D. Whitaker, MD, Health Sciences Centre, The University of Calgary, Calgary, Alberta (Canada)

Neurogenic Prolactin Release: Effects of Mastectomy and Thoracotomy

V.S. Herman and W.J. Kalk

Endocrinology Unit, Department of Medicine, University of the Witwatersrand Medical School, Hillbrow, Johannesburg

Galactorrhoea is an uncommon but well-documented sequel of chest surgery and injuries to the chest wall. It has been reported in association with trauma and burns of the anterior chest, thoracotomy, mastectomy and following herpes zoster of thoracic dermatomes [*Salkin and Davis*, 1949; *Grossman et al.*, 1950; *Grimm*, 1955; *Barnes*, 1966; *Berger et al.*, 1966; *Aufses*, 1955]. *Morley et al.* [1977] reported a patient who developed persistent galactorrhoea following extensive burns to the anterior chest wall and breasts; serum prolactin levels were elevated in this patient, but were normalized by intercostal nerve block. This evidence for neurogenically stimulated prolactin release and reports of hyperprolactinaemia following mastectomy in some 40% of patients [*Willis et al.*, 1977] prompted the investigation of the acute effects of chest wall surgery on serum prolactin levels in patients undergoing mastectomy and thoracotomy.

Patients and Methods

Three groups of patients were studied: 10 patients aged 43–59 years who underwent mastectomy for breast carcinoma; 10 patients aged 15–45 years who had left-sided anterolateral thoracotomy operations for closed mitral valvotomy; 7 patients who had lower abdominal laparotomy operations for gynaecological procedures served as controls. No patient was receiving any medication, and thyroid and renal function were normal. In all subjects post-operative analgesia was carried out with intramuscular pethidine for up to 72 h and thereafter with paracetamol.

Single blood samples were taken from each patient the day before the operation and daily for 5 consecutive days postoperatively, between 0900 and 1100 hours. In the mastectomy group, 5 patients were studied 4 and 24 weeks after the operation. Samples were assayed for prolactin using a commercial homologous radio-immunoassay kit [*Hwang et al.*, 1971]. The intrassay and interassay coefficients of variation were 6 and 10%, respectively. Paired and unpaired t-tests were used in statistical analyses.

Results

Preoperative serum prolactin levels were normal in every patient, and the mean ($\pm$ SEM) values in each group were similar: 7.1 ± 1.3, 6.1 ± 1.6 and 4.3 ± 1.3 ng/ml, in the mastectomy, thoracotomy and laparotomy patients, respectively.

In the *laparotomy* patients serum prolactin levels did not alter consistently or significantly. Only 1 patient had an isolated high level of 65 ng/ml, on the 4th postoperative day.

In the *thoracotomy* patients, prolactin levels rose in 3 patients and in the others remained in the normal range; mean prolactin levels on days 1 and 2 were 16.8 ± 6.3 and 27.2 ± 12.7 ng/ml, respectively. The peak on the second postoperative day was contributed to by high levels in only 2 patients, with individual values of 80 and 117 ng/ml. By days 4 and 5, mean serum prolactin had fallen to 10 ng/ml. 1 patient, however, had progressively increasing levels postoperatively. Her prolactin level on day 5 was 52 ng/ml.

In the *mastectomy* group, there was a consistent rise in prolactin on day 1 in all patients from a mean of 7.1 ± 1.3 to 16.0 ± 3.3 ng/ml ($p<0.01$); levels were supranormal in 5 subjects. Prolactin concentrations continued to rise to a mean peak of 35.6 ± 6.6 ng/ml on day 5 ($p<0.005$) and was above normal in 9. On day 5, mean prolactin levels in mastectomy patients were significantly greater than in the thoracotomy group – 10.7 ± 5.3 ng/ml ($p<0.05$) – and the laparotomy patients – 9.3 ± 3.8 ng/ml ($p<0.025$). Serum prolactin was measured in 4 patients 1 month after mastectomy: the levels remained elevated in all, ranging from 20.8 to 88.4 ng/ml, with a mean value of 46.8 ng/ml. Levels were measured in 5 subjects 6 months after mastectomy: in 4, concentrations were within the normal range (1.2–14.3 ng/ml), but in 1, the level remained elevated at 34.8 ng/ml.

Discussion

Mastectomy resulted in consistent, progressive elevations of serum prolactin levels in all patients; peak concentrations were reached 5 days after the operation in 9 of the 10 subjects, and remained above the normal range for at least 4 weeks in the subjects evaluated. This operation damages the whole breast area, including the rich nerve supply to the nipple. Since intercostal nerve block lowers the raised serum prolactin levels associated with chest injury [Morley et al., 1977], it seems likely that the extensive surgical trauma resulting from the removal of a breast involves the nerves of the suckling reflex, and stimulates pituitary prolactin release via this reflex. The progressive rise in prolactin concentrations in the immediate postoperative days may have been caused by increasing degrees of neural stimulation as tissues began to heal.

In contrast to mastectomy, anterolateral thoracotomy operations cause a lesser degree of chest wall injury and involve only one intercostal nerve at some distance from the nipple area. Thus, thoracotomy exerted an inconsistent effect on prolactin levels. 2 subjects, however, had transiently elevated concentrations on the second postoperative day; a third subject followed the pattern of the mastectomy patients, with a rise to maximum concentrations on day 5. The remaining 7 thoracotomy patients followed the pattern of the laparotomy patients who served as the control group and in whom prolactin levels generally did not rise above the normal range.

In all patients studied we hoped to miss the expected intra-operative prolactin peak which usually returns to baseline within 24 h of operation [Noel et al., 1972]. The slight, unsustained prolactin fluctuations found in some of the laparotomy and thoracotomy patients could be attributed to continuing stress from the operation. In conclusion, we suggest that injury to the nerve supply of the breast may result in hyperprolactinaemia on the basis of peripheral neurogenic stimulation.

References

Aufses, A.H.: Abnormal lactation following radical mastectomy. N.Y. St. J. Med. 55: 1914–1915 (1955).
Barnes, A.B.: Diagnosis and treatment of abnormal breast secretions. New Engl. J. Med. 275: 1124–1127 (1966).
Berger, R.L.; Joison, J., and Braverman, L.: Lactation after incision of the thoracic cage. New Engl. J. Med. 274: 1493–1495 (1966).
Grimm, E.G.: Non-puerperal galactorrhoea with case reports. Q. Bull. Northwestern Univ. Med. School 29: 350–354 (1955).
Grossman, S.; Buchber, A.S.; Brecher, E., and Hallinger, E.M.: Idiopathic lactation following thoracoplasty. J. clin. Endocr. Metab. 10: 729–734 (1950).
Hwang, P.; Guyda, H.J., and Friesen, H.G.: Radioimmunoassay for human prolactin. Proc. natn. Acad. Sci. USA 68: 1902–1906 (1971).

Morley, J.E.; Dawson, M.; Hodkinson, J., and Kalk, W.J.: Galactorrhoea and hyperprolactinaemia associated with chest wall injury. J. clin. Endocr. Metab. *45:* 931–935 (1977).

Noel, G.L.; Suh, H.K.; Stone, J.G., and Frantz, A.G.: Human prolactin and growth hormone release during surgery and other conditions of stress. J. clin. Endocr. Metab. *35:* 840–851 (1972).

Salkin, D. and Davis, E.W.: Lactation following thoracoplasty and pneumonectomy. J. thorac. Surg. *18:* 580–590 (1949).

Willis, K.J.; London, D.R.; Ward, N.W.C.; Butt, W.R.; Lynch, S.S., and Rudd, B.T.: Recurrent breast cancer treated with the anti-oestrogen tamoxifen: correlation between hormonal changes and clinical course. Br. med. J. *i:* 425–428 (1977).

V.S. Herman, MD, Endocrinology Unit, Department of Medicine, University of the Witwatersrand Medical School, Hospital Street, Hillbrow, Johannesburg 2001 (South Africa)

Discussion

Adler: I should just like to mention a patient who came to my attention recently. I wonder if anyone else has had a similar experience. This woman had had the onset of galactorrhoea several months after a bilateral augmentation mammoplasty. Although we searched the plastic surgery literature, they claim that this does not happen. She had a slightly elevated prolactin level, in the 25–30 ng/ml range, and no other obvious cause of her galactorrhoea. Has anyone had experience of this particular kind of chest trauma?

Bern: Are there any data from experimental animals on what happens when you cut the nerve from the nipple area?

Herman: If you cut the nerve in animals which are already lactating, it does not interfere with the galactorrhoea. If the nipple is stimulated after nerve section, oxytocin will not be released and hence milk 'let down' will not occur. Milk production, once established, is not dependent on continued prolactin secretion.

Buttle: Transplanted mammary glands of goats to the neck area can lactate to the same extent as a gland left *in situ*. In fact, these glands do regenerate a sympathetic nervous system over a period of time, and prolactin levels are unaffected.

Herman: We have carried out a retrospective study in which we looked at 53 patients who had undergone mastectomy. Prolactin levels were supranormal in 34%. We could find no statistically significant relationships between serum prolactin levels and the extent of surgery, menopausal status, or period postoperation.

Autoregulation of Prolactin Secretion

Stephen J. Judd

Department of Medicine, Flinders Medical Centre, and Centre for Neuroscience, Flinders University, Bedford Park, South Australia

Numerous studies in the experimental animal have demonstrated that physiologically [*Fuxe et al.*, 1969] or pharmacologically induced [*Eikenburg et al.*, 1977] hyperprolactinaemia is followed by an increased activity of tubero-infundibular dopamine neurons (TIDA) and an increased release of dopamine (DA) into the portal vessels [*Cramer et al.*, 1979]. This effect of prolactin (PRL) appears to be a direct effect and quite specific for TIDA neurons [*Perkins et al.*, 1979].

Persistence of hyperprolactinaemia in women implies a breakdown in normal autoregulation either as a result of an inability of TIDA neurons to respond to an increase in PRL (afferent arm of feedback loop) or because the hypersecreting lactotropes are not able to respond to increased portal blood levels of dopamine (efferent arm).

We have previously reported that TIDA neurons inhibit the release of luteinising hormone (LH) as well as PRL [*Judd et al.*, 1978]. In this study, we investigated the ability of TIDA neurons to respond to hyperprolactinaemia by examining the LH response to metoclopramide (a dopamine receptor antagonist) in hyperprolactinaemic women.

Materials and Methods

6 women with persistent hyperprolactinaemia and abnormal pituitary tomograms suggesting microadenoma formation were examined; 6 normoprolactinaemic women with chronic anovulation and comparable age and weight were studied in the same way as controls.

After a baseline period of 2 h each woman received luteinising hormone-releasing factor (LRF: 10 μg i.v.) and 2 h later metoclopramide (MCP: 10 mg i.v.). Serial blood samples were collected at 15-min intervals for a total of 6 h. Samples were assayed for LH, FSH and PRL, using materials supplied by NIAMDD (NIH, Bethesda, Md.).

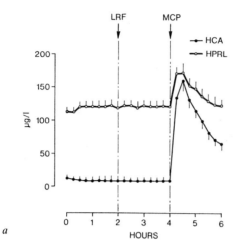
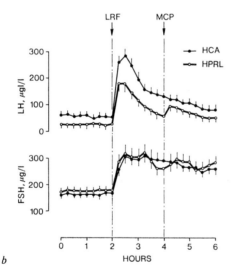

Fig. 1. Serum PRL (a) and LH and FSH (b) responses to LRF (10 μg i.v.) and metoclopramide (MCP; 10 mg i.v.) in 6 hyperprolactinaemic women (HPRL) and 6 women with normoprolactinaemic chronic anovulation (HCA).

Results

Basal serum levels of LH were lower in hyperprolactinaemic women (27.0 ± 2.9 µg/l) than in the control group (51.2 ± 2.0 µg/l; p < 0.001) and the maximum response to LRF stimulation was reduced (181 ± 30 vs 271 ± 60 µg/l). There was no significant difference in the basal FSH levels in the two groups or their response to LRF stimulation.

Following MCP there was an increase in serum LH in the hyperprolactinaemic woman but not the control group (fig. 1). This increase was small (57 ± 9 to 97 ± 13 µg/l, p < 0.01; paired t-test) but was consistent in each women studied. Furthermore, the duration and pattern of response was similar to that seen with LRF.

Basal serum prolactin levels in the hyperprolactinaemic women increased slightly from their elevated basal level (126 ± 12 µg/l to 167 ± 15 µg/l). There was a much greater increase in the control group (10.1 ± 0.5 to 146 ± 14 µg/l) (fig. 1).

Discussion

The clear LH response to MCP which is seen only in the hyperprolactinaemic anovulatory group is consistent with the view that TIDA neurons actively inhibit LRF secretion in hyperprolactinaemic women [*Quigley et al.,* 1979]. It is not known whether there is 'specialization' in TIDA neuronal activity but these findings indicate that the TIDA neurons controlling LRF release are able to perceive and respond to increasing PRL levels.

Conversely, the continued elevation in serum PRL and the reduced PRL response to metoclopramide indicate that the lactotrope does not receive, or does not respond to this increase in TIDA activity. There is some evidence which indicates that this may be due to a defect in the TIDA neurons [*Fine and Frohman,* 1978; *Van Loon,* 1978; *Müller et al.,* 1978]. If this is so, then there must be specific TIDA neurons controlling PRL secretion. Indeed, considering the abundance of neurotransmitters, releasing factors and brain peptides that are known to accumulate in the median eminence and its access to the peripheral circulation, it would be surprising if some degree of specialisation does not occur. If a specific TIDA neuron defect does occur, then it must be reversible considering the well established effectiveness of total resection of a prolactinoma and the restoration of normal patterns of PRL response after surgery [*Barbarino et al.,* 1978].

At the pituitary level, hyperprolactinaemia may develop because of an inability of hypersecreting lactotropes to perceive or respond to an increased TIDA activity. This could be explained by the sequestration of a nest of lactotropes outside the influence of the portal blood [*Porter et al.*, 1978] or due to an abnormality in dopamine receptor sensitivity [*Malarkey et al.*, 1977]. However, the prompt response of hyperprolactinaemic women to bromocriptine and dopamine does not support a receptor problem [*Evans et al.*, 1980].

Ideally, to resolve this issue, it would be necessary to measure local concentrations of dopamine at the level of the lactotrope in hyperprolactinaemic women. Meanwhile, local dopamine concentration in the pituitary can be deduced by studying the levels and responses of other dopamine-controlled pituitary hormones. Moreover, serial study of PRL response to metoclopramide in hyperprolactinaemic women may demonstrate a changing pattern of autoregulation of PRL and provide important information concerning the pathophysiology of this disorder.

Acknowledgements

I am grateful to *Craig Roeger* and *Roslyn Collingwood* for their technical assistance and to *Ingrid Muenstermann* for typing this manuscript. Materials for radio-immunoassay were kindly provided by NIAMDD, NIH, Bethesda. This work is supported by the National Health and Medical Research Council of Australia.

References

Barbarino, A.; De Marinis, L.; Maira, G.; Menini, E., and Quile, C.: Serum prolactin response to thyrotropin-releasing hormone and metoclopramide in patients with prolactin-secreting tumours before and after transsphenoidal surgery. J. clin. Endocr. Metab. 47: 1148–1151 (1978).

Cramer, O.N.; Parker, C.R., and Porter, J.C.: Secretion of dopamine into hypophysial portal blood by rats bearing prolactin-secreting tumours or ectopic pituitary glands. Endocrinology 105: 636–640 (1979).

Eikenburg, D.C.; Ravits, A.J.; Gudelsky, G.A., and Moore, K.E.: Effects of estrogen on prolactin and tubero-infundibular dopaminergic neurons. J. Neural. Trans. 40: 235–244 (1977).

Evans, W.S.; Rogol, A.D.; MacLeod, R.M., and Thorner, M.O.: Dopaminergic mechanisms and luteinising hormone secretion. I. Acute administration of the dopamine agonist bromocriptine does not inhibit luteinising hormone release in hyperprolactinaemic women. J. clin. Endocr. Metab. 50: 103–107 (1980).

Fine, S.A. and Frohman, L.A.: Loss of central nervous system component of dopaminergic inhibition of prolactin secretion in patients with prolactin secreting pituitary tumours. J. clin. Invest. *61:* 973–980 (1978).

Fuxe, K.; Hökfelt, A., and Nilsson, O.: Factors involved in the control of the activity of tubero-infundibular dopamine neurons during pregnancy and lactation. Neuroendocrinology *5:* 257–270 (1969).

Judd, S.J.; Rakoff, J.S., and Yen, S.S.C.: Inhibition of gonadotropin and prolactin release by dopamine: effect of endogenous estradiol levels. J. clin. Endocr. Metab. *47:* 494–498 (1978).

Malarkey, W.B.; Groshong, J.C., and Milo, E.E.: Defective dopaminergic regulation of prolactin secretion in a rat pituitary tumour cell line. Nature, Lond. *266:* 640–641 (1977).

Müller, E.E.; Genazzani, A.R., and Murru, S.: Nomifensine: diagnostic test in hyperprolactinaemic states. J. clin. Endocr. Metab. *47:* 1352–1357 (1978).

Perkins, N.A.; Westfall, T.C.; Paul, C.V.; MacLeod, R., and Rogol, A.D.: Effect of prolactin on dopamine synthesis in medial basal hypothalamus: evidence for a short loop feedback. Brain Res. *160:* 431–444 (1979).

Porter, J.C.; Barnes, A.; Cramer, O.M., and Parker, C.R.: Hypothalamic peptide and catecholamine secretion: roles for portal and retrograde blood flow in the pituitary stalk in the release of hypothalamic dopamine and pituitary prolactin and LH. Clin. Obstet. Gynaec. *5:* 271–282 (1978).

Quigley, M.E.; Judd, S.J.; Gilliland, G.B., and Yen, S.S.C.: Effects of a dopamine antagonist on the release of gonadotropin and prolactin in normal women and women with hyperprolactinaemic anovulation. J. clin. Endocr. Metab. *48:* 718–720 (1979).

Van Loon, G.R.: A defect in catecholamine neurons in patients with prolactin-secreting pituitary adenoma. Lancet *ii:* 868–871 (1978).

S.J. Judd, MD, Department of Medicine, Flinders Medical Centre, and
Centre for Neuroscience, Flinders University, Bedford Park,
South Australia 5042 (Australia)

General Discussion On the Lactotrope: Stimulation and Regulation of Its Activity
Chairmen: *J. Neill and S.J. Judd*

Buckman: In order to exclude factors known to alter prolactin, such as diurnal variation, stress, spontaneous fluctuations, we elected to do a study in which we measured serum prolactin at 15-min intervals for 5 h during the early follicular phase of the cycle, the low-estrogen phase, and again at mid-cycle, the high-estrogen phase. Although the differences are variable, the tendency was to secrete more prolactin during the high-estrogen phase, that is the mid-cycle phase; during the early follicular phase the mean was 8.1, and during the mid-cycle phase the mean was 10.7 ng/ml. Using paired Student's t test, this was highly significant at the $p < 0.001$ level. It is clear to us that there is a role for endogenous estrogens in the regulation of prolactin secretion. Unlike the rat, however, these changes are quantitatively much smaller in humans.

Thorner: I still think that the differences between the rat and the human are obvious from the data. Buckman has shown, as others, a very small change in prolactin, but there is not a proper peak of prolactin at mid-cycle in the human as there is in the rat pro-estrous, which is time-related. The LH and prolactin secretions are very different in the human and the rat.

Robyn: We have had the opportunity to investigate, with Aidar in the Ivory Coast, the menstrual cycle of the Mangabey monkey. Since it was not possible to validate an LH assay in these animals, we combined the cycles by aligning the maximal levels of estradiol, which is rather unusual. Following the estradiol peak by a day or two, there is a rise in the basal levels of serum prolactin. Then the values tend to decline during the progesterone peak, and to rise again at a later stage in the cycle. Thus, there is definite evidence that there is a relationship between estrogen secretion and prolactin secretion in the cycle.

Neill: I want to present some of our recent data on the effects of estrogens on prolactin secretion in the rhesus monkey and the interaction between that and dopamine. I hope to show you that the situation in this particular primate is somewhat different from the one that Thorner showed you in his overview.

Estrogen implants have been made into two different groups of animals: animals which have been stalk transected and also ovariectomized, and another group which were simply ovariectomized. The amount of estrogen implanted in these animals, with silastic capsules, produced levels to approximate that of mid-follicular phase (that is about 100 pg/ml), so that these were physiological amounts of estrogen. In response to the estradiol implants, there was absolutely no effect on the secretion of prolactin in the ovariectomized animals – the animals in which the hypothalamus and the pituitary remained intact. On the other hand, in the stalk-transected animals, which had elevated levels over the ovariectomized animals as a result of the transection, when estrogens were implanted there was a significant rise in prolactin secretion, of the order of 200%, which persisted as long as the transplant was in place. This suggests that when the hypothalamus is connected to the pituitary, it antagonizes the effect of estrogen to stimulate prolactin.

We wondered what might happen if we infused dopamine into these animals, and compared the time when they were not under the influence of estrogen and the time when they were under the influence of estrogen. We wanted to infuse amounts which are present in hypophyseal stalk blood into these animals. In order to select the dopamine infusion to be made, we have measured dopamine levels in hypophyseal stalk blood of the rhesus monkey and found an overall mean of 0.76 ng/ml, which is significantly higher than the dopamine levels in peripheral plasma, suggesting that dopamine is a secretion product of the hypothalamus of the monkey. We then selected an infusion rate of dopamine which would produce plasma levels equivalent to the amount present in stalk plasma, that is the 0.76 ng/ml. The effect of that infusion, which turned out to be 0.1 μg/min/kg body weight, was to produce no significant inhibition of prolactin in the stalk-transected animal; in fact, it required a pharmacologic dose – that is 1 μg/min/kg body weight – to significantly suppress prolactin secretion. However, and surprisingly, if we treat animals with estrogen and then infuse them with a physiological amount of dopamine, we get a very striking suppression of prolactin secretion. We conclude from these data that, although estrogen has the propensity to stimulate prolactin synthesis directly on the pituitary, it reinforces the inhibitory effect of dopamine on prolactin secretion. Judd and Yen have reported similar results in the human female. These data suggest that, in the primate, the roles of estrogens and dopamine on prolactin secretion are in direct contrast to their roles in the rat, in which estrogens and dopamine are mutually antagonistic.

Judd: When you infuse dopamine into the peripheral circulation do you in fact achieve comparable portal blood levels to those that you have previously measured?

Neill: We do not have that answer for the monkey, although in the rat we have shown very clearly that the amount which we infuse into the periphery is the amount which appears in the stalk blood, in animals treated with α-methylparatyrosine to totally block dopamine synthesis.

Robyn: I wonder if there are differences related to the mode of administration of the estrogens. As you know, there are other papers published which indicate that estrogens are raising prolactin levels in the monkey too. Do you think that the way estrogens are administered may influence the experiments? A silastic capsule is supposed to release constantly; is that the way that estrogens are secreted in the normal animal?

Neill: I do not know of anyone who has measured estrogens sufficiently frequently to specify whether there is pulsatile release of estradiol. I am not aware of any report, other than the one you made, about cyclic changes in prolactin levels in the monkey. Were you referring to your own data?

Robyn: There are other data showing changes in prolactin after estradiol treatment of ovariectomized rhesus monkeys.

Neill: Yes, Spies group has also reported that. You have to administer the estrogen over a prolonged period of time, and at much higher levels, I think. There is also Milmore who has given estradiol benzoate injections in very large amounts. He showed that, although those estrogen injections would not elevate plasma levels of prolactin or would only elevate them very slightly at the sort of levels that Buckman is showing – these very tiny increases to which I find it difficult to assign physiologic significance – the amount of prolactin in the pituitary gland goes very high. That fits well with what we have seen here, it seems to be a direct action of estrogen on the pituitary to cause synthesis of prolactin, which can be released in the absence of dopamine. But then, when dopamine is present, it synergized with estrogen to suppress prolactin release.

Kelly: It is important to point out to this group the work by Labrie et al. in our laboratory. They have shown, both *in vivo* and using pituitary cells in culture, the potent antidopaminergic action of estrogens: when estrogens are present in an incubation medium of pituitary cells, much more dopamine or dopamine agonist is required to induce an inhibition of prolactin secretion.

Jacobs: Following stalk section, one consequent event that may be of relevance in the interpretation of prolactin secretion is a significant hypothyroidism. Consideration of the role that hypothyroidism may play, either on receptors or other mechanisms of tissue sensitization, should not be forgotten.

Neill: In the human studies by Judd, the infusion of dopamine inhibited prolactin secretion much more when estrogen levels were high, and much less when estrogen levels were low. These women did not have thyroid problems. There is no doubt from the studies of Labrie that there is this antagonism between estrogen and dopamine, while in primates, including perhaps the rhesus monkey and the human female, there is this synergism between the two.

Shin: It is more difficult to suppress prolactin secretion with dopamine *in vitro* but in our hands, the *in vivo* situation appeared slightly different. It very much depends on the initial concentration of prolactin: If that is high, it is suppressed more by the same amount of dopamine.

Neill: Dopamine, infused in this low dose which we think is the physiologic dose, will suppress prolactin secretion in animals in the early follicular phase, whereas the same amount of dopamine does not inhibit prolactin secretion in the stalk-transected animal which has not been treated with estrogen. The stalk-transected animal has very high prolactin levels, and is insensitive to the dopamine infusion, whereas a follicular phase animal, which has low prolactin but which has estrogens is sensitive to the dopamine infusion.

I should now like to discuss some of our recent findings on the issue of whether changes in dopamine secretion into hypophyseal stalk blood can account for changes in prolactin secretion, specifically in the lactating female rat. We anesthetized lactating female rats with urethane, isolated the mammary nerve, and stimulated it electrically: that causes a very nice rise in prolactin secretion, which is very like that seen after suckling. Control animals, in which the nerve was isolated but not stimulated, showed no increase in prolactin secretion. We collected stalk blood during the same period of time. In the control group there is no significant change of dopamine over time; in contrast, in the animals in which the mammary nerve was stimulated, there is a significant 20% drop during the first 15 min of stalk blood collection period; in the second 15 min, there is a return to the baseline, and then there is a significant increase in dopamine secretion into the hypophyseal stalk blood. Since we were worried somewhat about what the collection of stalk blood might mean, we have developed a new method of measuring dopamine with an electrochemical probe which one can implant directly into the brain to measure catecholamine release. We insert the probe directly into the median eminence, presumably into the region from which the portal capillaries originate.

In quite a number of rats, there was a significant increase in dopamine or at least in electrochemical current during mammary nerve stimulation, which causes an increase in prolactin secretion. After mammary nerve stimulation was stopped, we began to get major oscillations in the electrochemical current, presumably representing changes in dopamine release at the synaptic clefts, or at least these modified synaptic clefts on which the dopaminergic nerves end in the portal capillaries. There is a very good correspondence between these data and what we measured in hypophyseal stalk blood, that is a brief and small decrease in dopamine secretion. What does such a small decrease do? Can it account for a major rise in prolactin secretion? Taking a group of rats treated with α-methylparatyrosine, to completely inhibit synthesis of dopamine and the secretion of dopamine into the hypophyseal stalk blood, they have very high PRL levels and, when you infuse dopamine again in physiologic concentrations there is a suppression, as we might expect, in prolactin levels. We then decrease the dopamine infusion by about 30%, which is about what we have seen overall in response to electrical stimulation of the mammary nerve, and there is a 1.5- to 2-fold rise in prolactin levels, which is not nearly sufficient to account for the rise in prolactin secretion which follows mammary nerve stimulation, which is of the order of a 5- to 10-fold increase. We think that this brief decrease is important, but by itself it certainly cannot account for the major rise in prolactin secretion which follows this simulated suckling stimulus, mammary nerve stimulation.

Wuttke: Do you think that the oscillations have anything to do with dopamine, may reflect already an increased dopamine turnover due to increased prolactin levels, or do they reflect anything else? These oscillations did not show in the levels you measured in the portal blood if I remember correctly. Do you have any information about norepinephrine or epinephrine?

Neill: We are collecting over 15-min periods, whereas with the electrochemical probe we are measuring at 1-min intervals: the 1-min spikes get averaged across a 15-min period. Nevertheless, if you add those spikes together, you do not see the kind of increase in hypophyseal stalk blood that we would have expected on the basis of the spikes. We think that the spikes may represent the effect of prolactin, which has begun to feed back. One of the great difficulties which we have had with the probe has been trying to decide its specificity: it is probably measuring three things in the median eminence: dopamine, one of the metabolites of dopamine DOPAC (dihydroxyphenylacetic acid), and ascorbate; the ascorbate does not change, only the dopamine and the DOPAC change. We have also had problems trying to prove that what we measure in the median eminence is a measure of the dopamine which is destined to reach the portal capillary. It may not. As you know, there is considerable evidence for another PIF, which may itself be dopamine-dependent. What we are measuring may be dopamine release at synaptic sites which is concerned with the release of PIF and which is not destined for pituitary portal blood; we may be measuring dopamine destined for pituitary portal blood; perhaps more likely, we may be measuring a mixture of the two.

Frantz: Do you think that some of the relatively small changes in stalk blood dopamine that you measured might be due to regional specificity of the capillaries within the stalk, that relatively great changes in a few capillaries that were going to particular areas within the hypothalamus might have been partly washed out by no change in secretion to other capillaries in the portal vessels?

Neill: It is certainly a possibility. That sort of thing is very difficult for us to know about. Work about the various pools of prolactin and about how these pools may be differentially regulated by different hypothalamic factors, may be relevant. I think that what happens is that dopamine decreases, allowing a prolactin pool 1, or phase 1, to move to phase 2, which is very sensitive to TRH. We have some very preliminary evidence in my laboratory that a very brief decrease in dopamine – in pituitaries being infused *in vitro* – will allow prolactin to move into a pool which is highly sensitive to TRH, whereas before it was not. It is beginning to be apparent that prolactin secretion may consist of several phases, each of which has its own hypothalamic control; if there are enough of these phases, perhaps we can account for those 16 hypothalamic factors which seem to regulate prolactin secretion.

Voogt: Measuring tyrosine hydroxylase as an index of dopaminergic activity in the median eminence, we also, after about 60 min, find an increase in tyrosine hydroxylase activity, but we do not find a fall at 5 or 10 min. We would agree that, at about 60 min, when prolactin levels have been elevated for some time, there is an increase in dopaminergic activity in the median eminence.

Greenwood: I am having a little difficulty this morning with the use of the term significant. I feel that we should try to reach consensus on what prolactin does and what is a significant amount physiologically. Then, perhaps, some of these changes could be classified simply as perturbations of a system and not true stimuli.

Neill: Indeed, we asked what change in dopamine would the pituitary consider significant, that is more important than what the computer considers significant. The pituitary does not consider 20% decrease to be very important, with respect to prolactin at least, while statistically by all standards it is a significant decrease.

The Prolactin Target Cell and Receptor
An Overview

Robert P.C. Shiu[1]

Department of Physiology, Faculty of Medicine,
University of Manitoba, Winnipeg, Man.

The diverse actions of prolactin were reviewed by *Bern and Nicoll* [6] and *Horrobin* [25] who compiled more than 80 reported actions of this hormone in species from teleost fish to mammals. In mammals, the mammary gland is undoubtedly one of the principal target organs of prolactin. It is the mammary gland where most of our present knowledge regarding the mechanism of action of prolactin has been derived from. In this tissue, prolactin can be a lactogenic and/or mitogenic hormone [12]. While the former role of prolactin is well defined, the latter, being studied mainly in relation to growth of neoplastic mammary tissues, is less so. We shall concern ourselves here with the mechanism of action of prolactin in the mammary epithelial cells and, to a lesser extent, in other cell types.

As is the case for many other polypeptide hormones, prolactin appears to initiate its action by interacting with a specific receptor on the cell surface of the mammary epithelial cell. This mode of action was first suggested by *Turkington* [69] who showed that prolactin covalently linked to Sepharose beads is biologically active in mouse mammary epithelial cells. He concluded from this result that prolactin initiates its effect by an action on the cell membrane because prolactin-Sepharose complexes presumably cannot enter the cell. Subsequent studies, however, suggest that this result might have been due to a slow release of prolactin from the Sepharose beads. However, autoradiographic and immunocytochemical studies [8, 40] demonstrated that ^{125}I-labeled and endogenous prolactin is localized on cell membranes of mammary epithelial cells. A great deal of information on prolactin receptors in the rabbit and rat mammary glands has accumulated in the last 6–7 years.

[1] Scholar, Medical Research Council of Canada.

The Prolactin Receptor

The prolactin receptors are enriched in plasma membrane fraction derived from pregnant and lactating rabbit mammary glands [61]. The same membrane fraction also exhibits highest specific activity of 5'-nucleotidase, confirming that receptors for prolactin are located on the cell membrane. Prolactin receptors are probably lipoprotein complexes because digestion with proteases and phospholipase C destroys the activity of receptors [61]. After solubilization by nonionic detergents such as Triton X-100, the prolactin receptor molecule has an apparent molecular weight of 220,000 daltons, a Stoke's radius of 50 Å and an isoelectric point (pI) of about 6.0–6.5 [62–64]. It binds to concanavalin A-sepharose column [*Ohgo and Friesen*, unpubl. observation] indicating that carbohydrate moieties are an intrinsic part of the receptor molecule. Thus, the prolactin receptor, by analogy with many other membrane proteins, would appear to be a glycoprotein with the hydrophobic portion embedded in the lipid bilayer of the plasma membrane and the hydrophilic portion that contains the carbohydrate moiety protruding outwards.

The membrane receptor sites for prolactin exhibit very high affinity for the hormone, the dissociation constant (K_d) being in the order of 10^{-10} M [61]. This shows that binding of prolactin to the receptors occurs readily at physiological concentrations of the hormone; half-saturation of the receptor sites occurs at a hormone concentration of about 7 ng/ml. The prolactin receptor sites in the rabbit mammary gland not only bind prolactin but also recognize other lactogenic hormones such as placental lactogens and human growth hormone but not nonprimate growth hormone which are not lactogenic in the rabbit [68].

In order to demonstrate unequivocally that the membrane receptor sites mediate the lactogenic response of prolactin in the mammary gland and that the binding of prolactin to membrane receptor indeed represents the first step in the action of hormone, we purified the receptor molecules. The membrane receptors were first solubilized by the nonionic detergent, Triton X-100. The soluble receptors were purified by affinity chromatography [62] and the purified receptors were used to raise antibodies in guinea pigs [65]. The antibodies block binding of prolactin to its receptors but have no effect on the binding of insulin to its receptors in the rabbit mammary gland [65]. Antireceptor antibodies are also able to block prolactin-mediated incorporation of [³H]-leucine into casein and transport of [¹⁴C]-aminoisobutyric acid, but are without effect on insulin-mediated events in explants of rabbit mammary glands maintained in organ culture [66]. These findings provide direct evidence for an obligatory functional role of a membrane receptor in mediating the action of a polypeptide hormone for the first time.

At this point, it is necessary to emphasize that prolactin receptors are widely distributed in many organs [42, 61]. Prolactin receptor activity is particularly high in the liver, kidney, adrenal and the gonadal tissues (prostate, testis and ovary). These nonmammary prolactin receptors possess prolactin-binding characteristics similar to that of the mammary gland. In view of the diverse action of prolactin, it is not surprising to find prolactin receptors in these organs. With the availability of antiprolactin receptor serum, we had the opportunity to further examine the properties of the prolactin receptors in these organs and to examine their biological significance. Hence, we found that antibodies to rabbit mammary prolactin receptors also inhibit the binding of prolactin, but not the binding of other hormones such as insulin and bovine growth hormone, to a variety of organs derived from a number of species. These organs include normal mammary gland and mammary gland tumors from mouse, rat and human; liver from rat and rabbit; ovary and prostate from rat [65]. These findings suggest that the immunological determinant of the receptor molecule is very similar for tissues derived from both sexes and from different species. Furthermore, when the antibodies were administered to rats *in vivo* during an estrus cycle, the most notable finding was an increase in the number of corpora lutea and an increase in serum prolactin level [11]. These *in vivo* results suggest that the antibodies block the luteolytic effect of prolactin in the ovary and that deficiency of prolactin is sensed by the animal, activating compensatory mechanisms to increase prolactin secretion in an attempt to overcome the deficient state.

Regulation of Prolactin Receptors

Since the prolactin receptors are obligatory for mediating some, if not all, of the actions of the hormone, it is possible that the receptors may be important in determining the sensitivity of the tissue to the hormone. If that is the case, it may be expected that physiological factors that affect tissue sensitivity to prolactin may also influence the activity of the receptor. Several studies revealed that this may be the case. Prolactin binding activity in lactating rat mammary gland is many times higher than that in tissues from pregnant animals [21, 23]. This increase is abolished when suckling by pups is prevented [10]. In addition, ergocornine (an inhibitor of prolactin secretion) also diminishes this postpartum increase in prolactin receptors [10]. These findings also suggest that the prolactin release caused by suckling increases prolactin receptor levels in the rat mammary gland. Postpartum increase in prolactin receptors is also observed in the rabbit mammary gland [16]. Administration of ovine prolactin induces prolactin receptors in the rabbit mammary, and this effect can be blocked by progesterone [17]. Using mouse mammary gland organ cultures, it was demonstrated that the thyroid hormone, triiodothyronine (T^3), stimulates prolactin receptors while it simultaneously potentiates the effect of prolactin on the induction of lactose synthetase [7]. These findings seem to extend the well-known inhibitory effect of progesterone and the synergistic effect of thyroid hormone on lactogenesis to the level of the prolactin receptors.

It should be emphasized that not only the prolactin receptors of the mammary gland but also prolactin receptors of other organs such as the liver, kidney, ovary and prostate are regulated by prolactin and other hormones. In general, they are sensitive to an androgen-estrogen balance. High level of testosterone suppresses prolactin receptor activity in rat liver but enhances that in rat ventral prostate [3]. Administration of estrogen or reduction of testosterone by castration results in the opposite [3, 43]. However, estrogen induction of hepatic prolactin receptors requires the presence of the pituitary gland [43]. This pituitary factor that is required for estrogen induction of hepatic prolactin receptors is thought to be prolactin because administration of high doses of prolactin partially restores the level of hepatic prolactin receptor in hypophysectomized female rats [9, 13, 44]. However, suppression of prolactin secretion by ergot alkaloids does not abolish the inductive effect of estrogen [3]. Further, prolactin injection does not increase hepatic receptors in normal or hypophysectomized male rats [1, 13]. In addition, castration leads to an increase in hepatic receptor despite the fact that castration lowers serum prolactin level [3]. These studies suggest that a pituitary factor(s) other than prolactin mediates the receptor induction process. Although a number of studies [2, 9] suggest growth hormone, adrenocorticotropic hormone and thyroid-stimulating hormone may also be important in regulation of prolactin receptors, there is no solid conclusion derived from these studies. It will not surprise me if a yet to be discovered 'hormone' in the pituitary is responsible for mediating the induction of prolactin receptors.

Postreceptor Events

While major efforts have been spent in elucidating the hormone-binding characteristics of the prolactin receptors and the mechanisms by which the receptors are regulated, the fate of prolactin after it interacted with the cell surface receptor has received little attention. Postreceptor events leading to the manifestation of the prolactin effect are still very poorly understood. Nevertheless, some understanding on the events after the binding of prolactin to its receptor have been obtained. The majority of this knowledge is derived from studies using the mammary tissue, and the following discussion will therefore be confined to this tissue, bearing in mind that similar studies have to be carried out in other organs as soon as the biological significance of prolactin in these tissues becomes better defined.

In spite of the general belief that polypeptide hormones exert their effects without entering the target cells, there is increasing evidence to suggest that hormones do enter cells. The mechanism by which receptor-bound hormones enter the cells and the physiological significance of this phenomenon are still unresolved. Immunocytochemical [40] and autoradiographic [4, 27] studies suggest that prolactin is intracellularly localized in its target cells such as mammary epithelial cells and hepatocytes. To gain better insight into this phenomenon, I took advantage of my finding that human breast cancer and normal cells maintained in long-term tissue culture retain specific binding of prolactin [58]. These intact, living cells maintained *in vitro* may prove to be excellent models to investigate postreceptor events in the action of prolactin. It was found that in these human breast cancer cells, receptor-bound prolactin is quickly internalized, possibly via pinocytosis, and degraded in intracellular sites (possibly the lysosomes) [59, 60]. The internalization process can be inhibited by metabolic poisons such as 2,4-dinitrophenol and sodium azide. The internalized prolactin molecule is degraded to at least three small molecular weight species which are released into the incubation media. Degradation of prolactin can be prevented by the protease inhibitor N-α-p-tosyl-L-lysine chloromethylketone (TLCK), and the lysosomotropic agents ammonium chloride and chloroquine. When prolactin degradation is inhibited, specific binding and subsequent dissociation of intact prolactin is still observed, suggesting that hormone degradation is not responsible for binding and dissociation of prolactin. There are quantitative differences in the ability to bind and degrade prolactin among the cell lines although there is a good correlation between the number of prolactin receptor sites and prolactin degradative activity. This suggests that the prolactin molecule will be internalized and degraded only if it binds to the receptor site. Autoradiographical studies also reveal intracellular as well as membrane-associated ^{125}I-labeled human prolactin in the human breast cancer cells [Shiu, unpublished observation]. It should be re-emphasized that intracellular prolactin has been found in normal rat and rabbit mammary epithelial cells [4, 40]; prolactin internalization is not unique to human breast cancer cells in tissue culture. The physiological significance of prolactin internalization and degradation remain to be elucidated.

In the mammary gland, prolactin binding to cell surface receptors leads to the ultimate production of milk component (e.g. lactose and casein). The biological actions of prolactin in other mammalian organs such as liver, kidney, ovary and prostate are less well defined. Nevertheless, prolactin has been reported to stimulate somatomedin production [20] and ornithine decarboxylase activity in rat liver [45] and to affect kidney's ability to retain Na^+ ions [39]. The luteal action of prolactin in the ovary is well documented [74]. Prolactin can also stimulate prostatic growth [35] and the rate of uptake of testosterone by the prostate [19]. In spite of the many reported actions of prolactin, the mechanisms by which the biological effects in various organs brought about by prolactin are far from being understood. The following discussion will be devoted to what is known with regard to the mechanism of action of prolactin in the induction of milk components. It should be borne in mind that different mechanisms may exist for prolactin in its action on different organs. Even with the mammary gland, prolactin is known to stimulate not only milk protein synthesis but also ion transport, lipid metabolism, secretion of milk-borne immunoglobulin and possibly cell proliferation [see review in 67]. Different mechanisms may exist for these actions of prolactin.

As far as the lactogenic action of prolactin is concerned, we now know that binding of prolactin to its cell surface receptor is essential for the subsequent induction of milk protein synthesis [66]. Information regarding the molecular events that take place between the binding of prolactin to its receptor and the ultimate stimulation of milk protein synthesis remains fragmentary.

Mediators of Prolactin Activity

Unlike most polypeptide hormones, prolactin binding to cell membrane receptors in the mammary tissue does not seem to activate the membrane-bound enzyme, adenylate cyclase [31, 70]. Furthermore, exogenous adenosine 3′,5′-cyclic monophosphate (cAMP) or its analog N^6-2′-0-dibutyryl cAMP does not mimic the action of prolactin [31, 48]. Therefore, cAMP production does not seem to be the rate-limiting step. What then is the second messenger system, if any, for prolactin? A series of experiments performed by *Rillema* and co-workers [46–54] suggest that activation by prolactin of the membrane-associated enzyme, phospholipase A, is perhaps very closely coupled to prolactin binding to receptor sites in the mouse mammary gland. These investigators showed that phospholipase A and arachidonic acid mimic the effect of prolactin on stimulating RNA synthesis. Since the activation of phospholipase is a rate-limiting step in prostaglandin synthesis, the above findings suggest the possibility that some of the effects of prolactin are mediated by prostaglandins. This notion is supported by their finding that prostaglandins B_2, E_2 and $F_{2\alpha}$ stimulate RNA synthesis in mammary gland explants of mice in a prolactin-like manner. These effects are nonadditive to those observed with concentrations of prolactin which stimulate the response to a maximal degree. Indomethacin, an inhibitor of prostaglandin synthesis, blocks the stimulatory effect of prolactin. Interestingly, the effects of prolactin as well as that of prostaglandins $F_{2\alpha}$, B_2 and E_2 are abolished by exogenous dibutyryl cAMP or theophylline, an inhibitor of phosphodiesterase, which elevates the intracellular concentration of cAMP. Moreover, guanosine 3′,5′-cyclic monophosphate (cGMP) mimics the early action of prolactin on RNA synthesis. These observations are therefore compatible with the idea that the action of prolactin on RNA synthesis may be mediated by a reduced intracellular concentration of cAMP coupled to an elevated level of cGMP. This hypothesis is further supported by the observations that prostaglandin E_1 and $F_{1\alpha}$, which are known to stimulate adenylate cyclase activity and thus to increase cAMP concentration in rabbit and mouse mammary glands, are able to abolish the effect of prolactin.

Although prostaglandins stimulate RNA synthesis in a prolactin-like manner, they cannot mimic the effect of prolactin on casein synthesis [49]. Some other factors are therefore required in addition to prostaglandins. It was subsequently shown that a combination of polyamines (e.g. spermidine) and prostaglandins exhibit a prolactin-like effect in stimulating casein synthesis in mouse mammary explants; polyamines by themselves cannot substitute for prolactin [50, 53]. Methylglyoxal bis-guanylhydrazone (methyl GAG), an inhibitor of polyamine synthesis, abolishes the stimulation of casein synthesis by prolactin. Addition of exogenous spermidine rescues the effect of prolactin from methyl GAG inhibition. Further, prolactin stimulates the activity of ornithine decarboxylase, an enzyme involved in the pathway of polyamine biosynthesis (ornithine→putrescine) [41, 51]. It thus seems probable that the action of prolactin on milk protein synthesis in the mammary gland may be mediated, at least in part, by polyamines and prostaglandins.

Protein phosphorylation has been implicated as an intermediate step of prolactin action in the mammary gland [70]. In mouse mammary gland organ cultures, prolactin induces the appearance of protein kinases. The consequences of this is the stimulation of phosphorylation of proteins in the plasma membranes, followed by that of ribosomal and nuclear proteins [70].

It is now clear that the nuclear event that follows is the stimulation of transcription of messenger RNA for milk proteins (e.g., casein and α-lactalbumin) [5, 26, 32, 37]. Prolactin stimulates the rate of production as well as prolongs the half-life of mRNAs for casein [22], in contrast to the prolactin-like effects of prostaglandins and polyamines in the stimulation of casein synthesis [41, 49–51, 53]. These compounds have indeed no appreciable effect on the accumulation of casein mRNA [33]. This suggests that polyamines and prostaglandins mediate the action of prolactin in stimulating casein synthesis posttranscriptionally (these compounds may stimulate the translation of casein mRNA). Progesterone, which inhibits lactation and reduces the number of membrane prolactin receptors (discussed earlier), blocks the prolactin-mediated transcription of casein mRNA [22, 26, 32]. These findings lend further support to the notion that the molecular mechanism of action of prolactin is complex and a great deal of effort will be required to gain better insight into the phenomenon.

Prolactin and Mammary Tumors

The mitogenic role of prolactin in neoplastic mammary tissue has been the subject of many studies. I do not intend to discuss the evidence for a role of prolactin in the etiology of breast cancer. This area has been adequately reviewed [29, 36, 73]. Instead, recent studies on the interaction of prolactin with breast cancer cells will be highlighted. Prolactin receptors are present in variable quantities in carcinogen-induced rat mammary tumors [14, 15, 24, 28]. Rat mammary tumors which grow in response to prolactin administration possess higher levels of prolactin receptor than prolactin-unresponsive tumors after treatment with prolactin [24, 74]. However, if prolactin binding is assessed prior to prolactin treatment, there is no significant difference between prolactin-responding and nonresponding (autonomous) tumors [24]. On the other hand, those tumors with higher prolactin receptor content were likely to regress following suppression of prolactin by bromoergocryptine [24]. Although the exact relationship between prolactin receptor and prolactin dependent growth in rat mammary tumors remains to be established, prolactin is reported to affect many other functions in at least some receptor-positive cancer tissues. For example, *in vivo* prolactin injections increase the level of estrogen receptors in DMBA-induced rat mammary tumors [55, 71]. Prolactin receptors are present in a transplantable rat mammary tumor (R3230AC) and prolactin stimulates the content of translatable messenger RNA for casein (a differentiated function that is unique to the mammary gland) in these tumors [38]. Prolactin-treated cell cultures or organ cultures of rodent breast carcinomas respond with an increase in DNA synthesis [72].

Studies on the role of prolactin in the development of rodent mammary tumors have provided, in part, the impetus to explore the hypothesis that this hormone may be similarly involved in human breast tumorigenesis. However, evidence accumulated so far has not established the importance of prolactin, if any, on the development of the human disease. However, studies with human breast tumor biopsies maintained in organ culture using a pentose pathway histochemical effect suggest that some tumors show some form of prolactin dependency [57]. Further, human breast biopsies maintained in organ culture [72] or transplanted in athymic nude mice [34] respond to prolactin and placental lactogen with an increase in DNA synthesis. One of the recent advances in human breast cancer research has been the development of a considerable number of permanent cell lines derived from breast cancer specimens. The use of these human breast cancer cell lines maintained in long-term tissue culture has widened our knowledge in the hormonal control of this neoplastic state in man. Some of these cell lines contain receptors for a variety of hormones [18]. Their proliferation as well as cellular functions are affected by many hormones [18, 30]. Many of these cell lines are shown to contain variable levels of prolactin receptors [58]. These human breast cancer cell lines provided me with an opportunity to examine the characteristics of the human prolactin receptors. The prolactin binding characteristics of these human breast cancer cells are in many ways similar to that of animal breast tissues [58]. The use of these cell lines to study internalization of receptor-bound prolactin has been discussed earlier. Prolactin binding to these cultured human breast cancer cells has been studied by immunohistochemical techniques using soluble peroxidase-antiperoxidase complexes [56]. This technique also reveals that different cell lines bind different amounts of prolactin. But the most interesting finding is that within one cell, prolactin binding is heterogeneous. Some cells exhibit high binding while other cells showing low to complete absence of prolactin binding [56]. The significance of the heterogeneity in prolactin binding remains to be studied. It is sufficient to say that these studies are only the beginning; these human breast cancer cell lines maintained in long-term tissue culture could be used for future studies on the mechanism of action of prolactin and other humoral factors in the control of proliferation and functions of human breast cancers.

Conclusions

In summary, there is no doubt that increasing understanding of the mechanism of action of prolactin in its target cells has been achieved. However, many gaps in our knowledge of the mechanism of action of prolactin remain. The nature of the prolactin receptor has to be better understood. It is therefore necessary to purify the receptor to homogeneity in sufficient quantity such that its structure can be determined. The interactions between the prolactin receptor and other cellular components (e.g., membrane lipids, membrane enzymes, cytoskeleton) remain to be elucidated. Investigation of the events immediately following prolactin binding to receptor deserves high priority. The putative second messenger system(s) for prolactin remains to be delineated. We do not yet understand the molecular mechanisms by which prolactin and its intracellular messenger affect cellular activities (such as gene expression and enzyme activities) that may eventually lead to the prolactin effects. The mechanisms of action of prolactin in tissues other than the mammary gland deserve more attention. For example, we do not even understand the precise function of prolactin in the liver, adrenal and kidney. Finally, information is urgently needed concerning the mechanism by which prolactin affects cell proliferation and neoplasia, as in the case of developing mammary gland and growth of breast cancer.

References

1 Aragona, C.; Bohnet, H., and Friesen, H.G.: Prolactin binding sites in the male rat liver following castration. Endocrinology *99:* 1017–1022 (1976).
2 Aragona, C.; Bohnet, H.; Fang, V.S., and Friesen, H.G.: Induction of lactogenic

receptors. II. Studies on the liver of hypophysectomised male rats and on rats bearing a growth hormone secreting tumor. Endocr. Res. Commun. *3:* 199–208 (1976).

3 Aragona, C. and Friesen, H.G.: Specific prolactin binding sites in the prostate and testis of rats. Endocrinology *97:* 677–684 (1975).

4 Aubert, M.L.; Suard, Y.; Sizonenko, P.C., and Krachenbuhl, J.P.: Prolactin receptors: study with dispersed cells from rabbit mammary gland: in Program and Abstracts, 60th Annu. Meet. Endocrine Soc., abstract 19, p. 84 (1978).

5 Banerjee, M.R.; Terry, P.M.; Sakai, S.; Lin, F.K., and Ganguly, R.: Hormonal regulation of casein messenger RNA. In Vitro *14:* 128–139 (1978).

6 Bern, H.A. and Nicoll, C.S.: The comparative endocrinology of prolactin. Recent Prog. Horm. Res. *24:* 681–720 (1968).

7 Bhattacharya, A. and Vonderhaar, B.K.: Thyroid hormone regulation of prolactin binding to mouse mammary glands. Biochim. biophys. Res. Commun. *88:* 1405–1411 (1979).

8 Birkinshaw, M. and Falconer, I.R.: The localization of prolactin labelled with radioactive iodine in rabbit mammary tissue. J. Endocr. *55:* 323–334 (1972).

9 Bohnet, H.; Aragona, C., and Friesen, H.G.: Induction of lactogenic receptors. I. In the liver of hypophysectomised female rats. Endocr. Res. Commun. *3:* 187–198 (1976).

10 Bohnet, H.; Gomez, F., and Friesen, H.G.: Prolactin and estrogen binding sites in the mammary gland of the lactating and non-lactating rat. Endocrinology *101:* 1111–1121 (1977).

11 Bohnet, H.G.; Shiu, R.P.C.; Grinwich, D., and Friesen, H.G.: *In vivo* effects of antisera to prolactin receptors in female rats. Endocrinology *102:* 1657–1661 (1978).

12 Ceriani, R.L.: Hormones and other factors controlling growth in the mammary gland: a review. J. invest. Derm. *63:* 93–108 (1974).

13 Costlow, M.E.; Buschow, R.A., and McGuire, W.L.: Prolactin stimulation of prolactin receptors in rat liver. Life Sci. *17:* 1457–1461 (1975).

14 Costlow, M.E.; Buschow, R.A., and McGuire, W.L.: Prolactin receptors in 7,12-dimethylbenz-(a)-anthracene-induced mammary tumors following endocrine ablation. Cancer Res. *36:* 3941–3942 (1976).

15 DeSombre, E.R.; Kledzik, G.; Marshall, S., and Meites, J.: Estrogen and prolactin receptor concentration in rat mammary tumors and response to endocrine ablation. Cancer Res. *36:* 354–358 (1976).

16 Djiane, J.; Durand, P., and Kelly, P.A.: Evolution of prolactin receptors in rabbit mammary gland during pregnancy and lactation. Endocrinology *100:* 1348–1356 (1977).

17 Djiane, J. and Durand, P.: Prolactin-progesterone antagonism in self-regulation of prolactin receptors in the mammary gland. Nature *266:* 614–616 (1977).

18 Engel, L.W. and Young, N.A.: Human breast carcinoma cells in continuous culture: a review. Cancer Res. *38:* 4327–4339 (1978).

19 Farnsworth, W.E.: Prolactin and the prostate; In Boyns and Griffiths, Prolactin Carcinog., Proc. 4th Tenovus Wkshop, pp. 217–225 (Alpha Omega, Cardiff 1972).

20 Francis, M.J.O. and Hill, D.J.: Prolactin stimulated production of somatomedin by rat liver. Nature *255:* 167–168 (1970).

21 Frantz, W.L.; MacIndoe, J.H., and Turkington, R.W.: Prolactin receptors: characteristics of the particulate fraction binding activity. J. Endocr. *60:* 485–497 (1974).

22 Guyett, N.A.; Matusik, R.J., and Rosen, J.M.: Prolactin-mediated transcriptional and post-transcriptional control of casein gene expression. Cell *14:* 1013–1023 (1979).
23 Holcomb, H.H.; Costlow, M.E.; Buschow, R.A., and McGuire, W.L.: Prolactin binding in rat mammary gland during pregnancy and lactation. Biochim. biophys. Acta *428:* 104–112 (1976).
24 Holdaway, I.M. and Friesen, H.G.: Correlation between hormone binding and growth response of rat mammary tumor. Cancer Res. *36:* 1562–1567 (1976).
25 Horrobin, D.F.: Prolactin 6 (Eden Press, Montreal 1978).
26 Houdebine, L.M. and Gaye, P.: Regulation of casein synthesis in the rabbit mammary gland: titration of mRNA activity for casein under prolactin and progesterone treatments. Mol. cell. Endocrinol. *3:* 37–55 (1975).
27 Josefsberg, Z.; Posner, B.I.; Patel, B., and Bergeron, J.J.M.: The uptake of prolactin into female rat liver: concentration of intact hormone in the Golgi apparatus. J. biol. Chem. *254:* 209–214 (1979).
28 Kelly, P.A.; Bradley, C.; Shiu, R.P.C.; Meites, J., and Friesen, H.G.: Prolactin binding to rat mammary tumor tissue. Proc. Soc. exp. Biol. Med. *146:* 816–819 (1974).
29 Kim, V. and Furth, J.: The role of Prolactin in carcinogenesis. Vitams Horm. *34:* 107–136 (1976).
30 Lippman, M.E.; Osborne, C.K.; Knazek, R., and Young, N.: *In vitro* model systems for the study of hormone-dependent human breast cancer. New Engl. J. Med. *296:* 154–159 (1977).
31 Majumder, G.C. and Turkington, R.W.: Adenosine 3',5'-monophosphate-dependent and independent protein phosphokinase isoenzymes from mammary gland. J. biol. Chem. *246:* 2650–2657 (1971).
32 Matusik, R.J. and Rosen, J.M.: Prolactin induction of casein mRNA in organ culture: a model system for studying peptide hormone regulation of gene expression. J. biol. Chem. *253:* 2343–2347 (1978).
33 Matusik, R.J. and Rosen, J.M.: Possible mediator of prolactin in the regulation of casein gene expression. Endocrinology *106:* 252–259 (1980).
34 McManus, M.J.; Dembroske, S.E.; Pienkowski, M.M.; Anderson, J.J.; Mann, L.C.; Schuster, J.S.; Vollwiler, L.L., and Welsch, C.W.: Successful transplantation of human benign breast tumors into the athymic nude mouse and demonstration of enhanced DNA synthesis by human placental lactogen. Cancer Res. *38:* 2343–2349 (1978).
35 Moger, W.H. and Geschwind, I.I.: The action of prolactin on the sex accessory glands of the male rat. Proc. Soc. exp. Biol. Med. *141:* 1017–1021 (1972).
36 Nagasawa, H.: Prolactin and human breast cancer: a review. Eur. J. Cancer *15:* 267–279 (1979).
37 Nardacci, N.J.; Lee, J.W.C., and McGuire, W.L.: Differential regulation of α-lactalbumin and casein messenger RNA's in mammary tissue. Cancer Res. *38:* 2694–2699 (1978).
38 Nardacci, N.J. and McGuire, W.L.: Casein and α-lactalbumin messenger RNA in experimental breast cancer. Cancer Res. *27:* 1186–1190 (1977).
39 Nicoll, C.S.: Physiological actions of prolactin. Handbook Physiol., Section 7, Endocrinol. *4:* 253–292 (1975).

40 Nolin, J.M. and Witorsch, R.J.: Detection of endogenous immunoreactive prolactin in rat mammary epithelial cells during lactation. Endocrinology 99: 949–958 (1976).

41 Oka, T. and Perry, J.W.: Studies on regulatory factors of ornithine decarboxylase activity during development of mouse mammary epithelium in vitro. J. biol. Chem. 251: 1738–1744 (1976).

42 Posner, B.I.; Kelly, P.A.; Shiu, R.P.C., and Friesen, H.G.: Studies on insulin, growth hormone and prolactin binding, tissue distribution, species variation and characterization. Endocrinology 95: 521–531 (1974).

43 Posner, B.I.; Kelly, P.A., and Friesen, H.G.: Introduction of a lactogenic receptor in rat liver: influence of estrogen and the pituitary. Proc. natn. Acad. Sci. USA 71: 2407–2410 (1974).

44 Posner, B.I.; Kelly, P.A., and Friesen, H.G.: Prolactin receptors in rat liver: possible induction by prolactin. Science 188: 57–59 (1975).

45 Richards, J.F.: Ornithine decarboxylase activity in tissues of prolactin-treated rats. Biochem. biophys. Res. Commun. 63: 292–299 (1975).

46 Rillema, J.A.: Effect of arachidonic acid on RNA metabolism in mammary gland explants of mice. Prostaglandins 10: 307–312 (1975).

47 Rillema, J.A.: Possible role of prostaglandin $F_{2\alpha}$ in mediating effect of prolactin on RNA synthesis in mammary gland explants of mice. Nature 253: 466–467 (1975).

48 Rillema, J.A.: Cyclic nucleotides and the effect of prolactin on uridine incorporation into RNA in mammary gland explants of mice. Hormone metabol. Res. 7: 45–49 (1975).

49 Rillema, J.A.: Effect of prostaglandins on RNA and casein synthesis in mammary gland explants of mice. Endocrinology 99: 490–495 (1976).

50 Rillema, J.A.: Activation of casein synthesis by prostaglandins plus spermidine in mammary gland explants of mice. Biochem. biophys. Res. Commun. 70: 45–49 (1976).

51 Rillema, J.A.: Action of prolactin on ornithine decarboxylase activity in mammary gland explants of mice. Endocr. Res. Commun. 3: 297–305 (1976).

52 Rillema, J.A. and Anderson, L.D.: Phospholipase and the effect of prolactin on uridine incorporation into RNA in mammary gland explants in mice. Biochem. biophys. Acta 428: 819–824 (1976).

53 Rillema, J.A.; Linebaugh, B.E., and Mulder, J.A.: Regulation of casein synthesis by polyamines in mammary gland explants of mice. Endocrinology 100: 529–536 (1977).

54 Rillema, J.A. and Wild, E.A.: Prolactin activation of phospholipase A activity in membrane preparations from mammary gland. Endocrinology 100: 1219–1222 (1977).

55 Sasaki, G.H. and Leung, B.S.: On the mechanism of hormone action in 7,12-dimethylbenz(a)anthracene-induced mammary tumor. I. Prolactin and progesterone effects on estrogen receptor in vitro. Cancer 35: 645–651 (1975).

56 Salih, H.; Cheng, K.W., and Shiu, R.P.C.: Immunocytochemical demonstration of prolactin binding to human breast cancer cells in long term tissue culture; in Program and Abstracts, 61st Annu. Meet. Endocrine Soc., abstract 316, p. 151 (1979).

57 Salih, H.; Flax, H.; Brander, W., and Hobbs, J.R.: Prolactin dependence in human breast cancers. Lancet ii: 1103–1105 (1972).

58 Shiu, R.P.C.: Prolactin receptors in human breast cancer cells in long term tissue culture. Cancer Res. 39: 81–86 (1979).

59 Shiu, R.P.C.: Prolactin binding and processing by human breast cancer cells in long term tissue culture; In Program and Abstracts, 61st Annu. Meet. Endocrine Soc., abstract 820, p. 277 (1979).
60 Shiu, R.P.C.: Processing of prolactin by human breast cancer cells in long term tissue culture (submitted for publication, 1979).
61 Shiu, R.P.C. and Friesen, H.G.: Properties of a prolactin receptor from the rabbit mammary gland. Biochem. J. *140:* 301–311 (1974).
62 Shiu, R.P.C. and Friesen, H.G.: Solubilization and purification of a prolactin receptor from the rabbit mammary gland. J. biol. Chem. *249:* 7902–7911 (1974).
63 Shiu, R.P.C. and Friesen, H.G.: Prolactin receptors. Methods mol. Biol. *9:* 565–598 (1976).
64 Shiu, R.P.C. and Friesen, H.G.: Studies on the isolation and purification of prolactin receptors; in Beer, Jr. and Bassett, Cell membrane receptors for viruses, antigens and antibodies, polypeptide hormones and small molecules, pp. 105–117 (Raven Press, New York 1976).
65 Shiu, R.P.C. and Friesen, H.G.: Interaction of cell membrane prolactin receptor with its antibody. Biochem. J. *157:* 619–626 (1976).
66 Shiu, R.P.C. and Friesen, H.G.: Blockade of prolactin action by an antiserum to its receptors. Science *192:* 259–261 (1976).
67 Shiu, R.P.C. and Friesen, H.G.: Mechanism of action of prolactin in the control of mammary gland function. A. Rev. Physiol. *42:* 83–96 (1980).
68 Shiu, R.P.C.; Kelly, P.A., and Friesen, H.G.: Radioreceptor assay for prolactin and other lactogenic hormones. Science *180:* 968–971 (1973).
69 Turkington, R.W.: Stimulation of RNA synthesis in isolated mammary cells by insulin and prolactin bound to sepharose. Biochem. biophys. Res. Commun. *41:* 1362–1367 (1970).
70 Turkington, R.W.; Majumder, G.C.; Kadohama, N.; MacIndoe, J.H., and Frantz, W.L.: Hormonal regulation of gene expression in mammary cells. Recent Prog. Horm. Res. *29:* 417–449 (1973).
71 Vignon, F. and Rochefort, H.: Regulation of estrogen receptors in ovarian-dependent rat mammary tumors. I. Effects of castration and prolactin. Endocrinology *98:* 722–729 (1976).
72 Welsch, C.W.; Calaf de Iturri, G., and Brennan, M.J.: DNA synthesis of human, mouse and rat mammary carcinomas *in vitro:* influence of insulin and prolactin. Cancer *38:* 1272–1281 (1976).
73 Welsch, C.W. and Nagasawa, H.: Prolactin and murine mammary tumorigenesis: a review. Cancer Res. *37:* 951–963 (1977).
74 Wuttke, W. and Meites, J.: Luteolytic role of prolactin during the estrous cycle of the rat. Proc. Soc. exp. Biol. Med. *137:* 988–991 (1971).

R.P.C. Shiu, PhD, Department of Physiology, Faculty of Medicine,
University of Manitoba, Winnipeg, Manitoba R3E OW3 (Canada)

Discussion

Guyda: Have the antisera to prolactin receptors been tested on any pituitary cells in culture, to get to the question of whether prolactin feeds back on its own secretion?

Shiu: No, it has not been used that way, but perhaps it can be. We are in the process of trying to generate more antisera to the receptors.

Waters: With your protease inhibitors and your chloroquine, you could obviously inhibit the degradation. What happened to the release of the prolactin from the cells? Did you still get the same sort of release rates?

Shiu: In the presence of chloroquine, for example, the release of radioactive hormone is much slower, but it is still being released. Probably, some disassociation is taking place, and the relase of internalized hormone is possible.

Waters: You think that hormone is not actually going in?

Shiu: It is going in, because chloroquine does not affect internalization, although it inhibits degradation once it is internalized. Some could be inside, that is why the release rate of internalized hormone in the presence of chloroquine is much slower.

Bohnet: You showed that progesterone is inhibiting the prolactin receptor. We have looked at the lactating mammary gland of the rat, and we did not find that. On the other hand, if you treat lactating women with progesterone, you do not see any decrease in milk yield.

Shiu: That was the work done by *Djiane* in France. They used pseudopregnant rabbits for the experiments. I have not done that sort of experiment.

Bern: The point that worries me is that it is important to be able to prove that prolactin is a mitogen on normal target, normal cells – mammary gland, crop sac, or whatever. One wonders whether your lymphoma cell-line data could be nothing more than a genetic freak, that these cancer cells happened to be dependent on prolactin.

Shiu: These cells are in a cancerous state, so they are transformed cells. It is possible that this is a property unique to these transformed cells. To go back to your question, as to whether prolactin is mitogenic in normal target tissue, such as the mammary gland, there is increasing evidence to show that it is.

Ahren: One should be careful about saying, from these studies, that prolactin is necessary for luteolysis, since luteolysis is such a complex event with so many hormones involved.

Carter: It was very interesting to hear your comments on the effects of thyroid hormone on prolactin receptors, particularly as there is some old data showing that prolactin receptors are increased in the mammary glands of thyroidectomized rats.

Shiu: It is interesting that stimulation by T_3 is not dependent on protein synthesis. T_3 may stimulate prolactin receptors not by a nuclear receptor, but it may interact with cell membranes to disturb the membrane enough to expose more functional receptors.

Barkey: Regarding the slight rise in prolactin serum levels in the rats treated with antiserum, I was wondering, rather than some feedback mechanism, whether this could not just reflect an increase in the unbound prolactin circulating in the periphery.

Jacobs: We know that the prolactin receptor is a glycoprotein, and in one sense, if it is in the Golgi, as has been suggested, it may be there just because it is on its way to the cell surface. If so, it would obviously be on the interior of the Golgi membrane, and I am not sure if localization is sufficiently fine to tell whether or not it is on the inside of the membrane. I wondered whether ^{35}S is the best label to use for fluorography; high specific activity ^{14}C glucosamine, even through it is ^{14}C, or tritiated glucosamine could be used, or has it been used, for similar sorts of experiments, perhaps with better efficiency.

Shiu: In terms of the specific labelling of the receptor, the reason that I use ^{35}S methionine is that it is the hottest thing you can buy. I have tried ^{14}C fucose, glucosamine, and none of them work, basically because there is so little receptors; even using ^{35}S methionine, you hit the tissue with several hundred micro Curies of isotope and only several thousand counts are associated with the receptors. With ^{14}C coumpounds, it just would not be hot enough. The other approach is iodination of cell membranes.

Guyda: When the vesicles of Golgi are interrupted by freeze-thawing, the binding is greatly enhanced, so that the binding is on the interior.

Induction of Hepatic Prolactin-Binding Sites in the Male Rat Liver: Role of Prolactin and Prostaglandins

R.J. Barkey, M. Lahav, J. Shani, T. Amit, M.G.H. Youdim and D. Barzilai[1]

Department of Pharmacology, Faculty of Medicine, Technion-Israel Institute of Technology, Haifa; Institute of Endocrinology, Rambam University Hospital, Haifa, and Department of Pharmacology, School of Pharmacy, The Hebrew University, Jerusalem

The role of prolactin (PRL) in the induction of its own receptor in the liver was always controversial. PRL has been shown to directly stimulate the level of its own receptor using hypophysectomized female rats bearing a renal pituitary transplant [17]. However, until recently, the success of PRL injections to effectively maintain PRL-binding sites has been less convincing since only partial restoration of PRL binding was demonstrated in one study [6], while other groups failed to show significant stimulation with injected PRL [1, 16]. The problem was solved by the use of a mixture of bovine PRL with polyvinylpyrrolidone (PVP) to achieve sustained blood levels of the hormone and thus restore fully PRL binding in the liver of hypophysectomized female rats [15].

We were interested in the mechanism by which PRL induces its own receptor. In the mammary gland, PRL has been shown to exert its biological action by influencing prostaglandin (PG) synthesis and it has been suggested that PGs may more generally mediate PRL action [for review see 10]. Using the adult male rat and PRL administered with PVP, we have studied the kinetics of the induction of the hepatic PRL receptor and the role of PGs in this action of PRL.

Methods

The methods for the preparation of the tissue homogenates, iodination and purification of PRL and testing of the binding activity of ovine PRL (^{125}I-oPRL) have been previously reported in detail [3, 4].

[1] D.B. is an established investigator of the Chief Scientist's Bureau, Israel Ministry of Health.

Injection of Hormones and Drugs

Unless otherwise indicated, injections were made s.c. using a volume of 0.5 ml/kg and administered twice daily, at 9:00 a.m. and at 5:00 p.m., for 6 days and once only on the morning of the 7th day (i.e. 6½ days, b.i.d.); the rats were killed 48 h after the last injection. A 20% w/w stock solution of PVP (av. molec. wt 360,000; from Fluka AG, Switzerland) was prepared by mixing one part by weight of PVP with 5 parts of 0.9% saline. This is the highest concentration of PVP of this molecular weight that we could prepare. When used as vehicle, this 20% solution was mixed with an equal volume of the drug or hormone solution in saline, to give a final PVP concentration of 10%. oPRL (P-S12) was dissolved in a minimum amount of 0.1% $NaHCO_3$ and diluted with saline; when required for injection this solution was mixed with an equal volume of PVP stock solution and thus kept at 5°C for up to 1 week. $PGF_{2\alpha}$ (tromethamine salt) was dissolved in saline and then mixed with an equal volume of 20% PVP solution, to a final concentration of 1.25 mg/ml and stored at 5°C for up to several weeks. PGE_1 and PGE_2 were dissolved in absolute ethanol and the 20-fold concentrated stock solutions stored at –20°C; daily, this was diluted 20-fold with 10% PVP in saline, to a final concentration of 1 mg/ml and kept at 5°C. Indomethacin was directly suspended in 10% PVP in saline and the suspension was stable for at least 1 h. Bromocryptine was dissolved in a small amount of absolute ethanol, then diluted with saline and, as such, was administered once/day for 6 days, at a dose of 2 mg/kg, the rats being killed on day 7. In one experiment, for *in vivo* 'desaturation' of PRL-binding sites [8], bromocryptine was injected, at a dose of 3 mg/kg, 3 times during 36 h following the last PRL injection. Anti-rPRL was raised in rabbits, against rPRL [23] and the antiserum, or normal rabbit serum (NRS) were injected i.p.; the injection schedule was 1 ml on day 1, 0.5 ml on day 3 and 0.5 ml on day 5, the rats being killed on day 7. Statistical analysis of the results was based on Student's two-tailed t-test, comparing experimental group means ($\pm$) and estimated SEM for significance.

Results

The effectiveness of the use of PVP 10% in saline, as a vehicle for PRL administration, is shown in figure 1. Twice daily injection, for 6½ days, of 500 µg/kg PRL dissolved in the PVP-containing vehicle, resulted in a 26-fold increase in the specific binding of ^{125}I-oPRL to the liver homogenate, compared with the vehicle control, while the same dosage of PRL, dissolved in saline only caused a 6-fold increase. Furthermore, in the group without PVP, the SEM was $\pm$ 75% of the mean, such that there was no statistically significant difference from the control, whereas the use of PVP reduced the SEM to $\pm$ 13% of the mean. In order to eliminate the possibility of masking of PRL receptor-sites by endogenous PRL, a third group of rats was administered bromocryptine acutely at the end of the PRL treatment [8]. Not only did this acute treatment not reveal any receptors masked by endogenous hormone, but it resulted in a somewhat lower increase in PRL binding

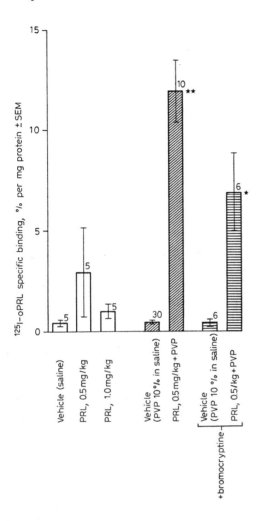

Fig. 1. Vehicle only, or PRL, either dissolved in saline or mixed with PVP 10% in saline were injected s.c. for 6½ days b.i.d. and the rats killed 48 h after the last injection. In one experiment, three bromocryptine injections (3 mg/kg, s.c.) were administered during 36 h, starting 12 h after the last PRL injection, and the rats killed 12 h after the last bromocryptine injection. The pooled number of rats, from one or more experiments, is indicated at the bottom of each column, the height of which indicates the mean ± SEM specific binding of ^{125}I-oPRL to the liver homogenate. Significance of the binding values of the PRL-treated rats vs those of the appropriate vehicle-treated controls. *p<0.01; **p<0.001.

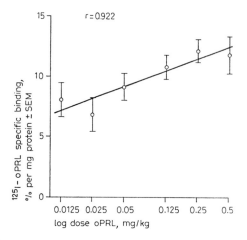

Fig. 2. The relationship between the dose of PRL (12.5–500 µg/kg) administered as described under figure 1 and the PRL-specific binding in the liver was tested for correlation by linear regression analysis. The pooled number of rats, from up to three separate experiments, ranged from 5 to 14 rats for each dose level.

by the exogenous PRL, possibly reflecting a peripheral effect of the bromocryptine on PRL uptake [14].

Figure 2 demonstrates the positive correlation (r = 0.922) existing between the dose of PRL administered and the level of PRL binding measured in the liver homogenate, for PRL doses of 12.5–500 µg/kg administered with PVP, twice daily for 6½ days; even a dose of 12.5 µg/kg, i.e. as low as 2.5 µg/rat, caused an 18-fold increase in PRL binding. The induction of the PRL receptor in the rat liver was also dependent upon the duration of the PRL administration *in vivo* (fig. 3). Only after 4½ days of PRL injection was there a statistically significant low increase in PRL binding, while a much larger increase was obtained after 6½ days. That this increase is a result of an induction in the number (N) of binding sites and not of a change in the affinity (K_a) of the receptor for PRL is shown by the data of the *Scatchard* [21] analysis in table I.

In an effort to understand the mechanism involved in this PRL induction of its own receptor, we first evaluated the importance of endogenous PRL in maintaining the normally low receptor levels found in the male rat liver. Administration of bromocryptine or rabbit anti-rPRL antiserum for 6 days, to reduce the effective serum levels of PRL [7], had no effect on PRL binding in the liver (table II).

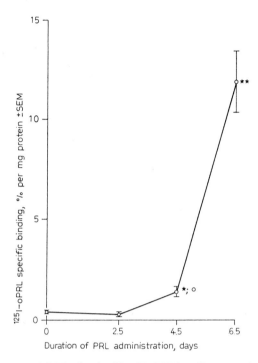

Fig. 3. PRL, dissolved in 10% PVP in saline, was administered at a dose of 0.5 mg/kg, s.c., b.i.d. for periods of 2½ days (6 rats), 4½ days (10 rats) and 6½ days (10 rats) and the rats were killed 48 h after the last injection. The number of animals indicated is the pooled number of 1 or 2 experiments, while there were 30 control rats, administered vehicle, from 6 experiments. ° $p<0.001$ vs control; * $p<0.01$ vs 2½ days; ** $p<0.001$ vs all groups.

Table I. Affinity (K_a) and concentration (N) (mean ± SEM) of PRL-binding sites in liver of male rats treated with PRL for various time intervals

Treatment	n	K_a ($\times 10^9\ M^{-1}$)	N, fmol/mg protein
Control	3	1.73 ± 0.26	0.73 ± 0.30
PRL 2½ days	3	1.48 ± 0.24	1.52 ± 0.34
PRL 4½ days	3	1.98 ± 0.23	4.65 ± 0.38**
PRL 6½ days	3	2.17 ± 0.66	153.11 ± 52.7*

* $p<0.05$; ** $p<0.01$ (versus vehicle-treated control).

Table II. Effect of reducing the effective serum PRL level on PRL-binding sites in the male rat liver

Treatment	n	Specific binding percent per mg protein ± SEM
Vehicle (control)	5	0.29 ± 0.09
Bromocryptine 2 mg/kg	5	0.36 ± 0.10
Anti-rPRL serum	5	0.26 ± 0.12

Table III. Effect of PGs on PRL binding in the male rat liver

Treatment		Days	n	Specific binding percent per mg protein ± SEM
Vehicle (control)		6½	30	0.45 ± 0.07
PGE_1	0.5 mg/kg	6½	5	0.48 ± 0.22
PGE_2	0.5 mg/kg	6½	5	0.94 ± 0.29
$PGF_{2\alpha}$	0.6 mg/kg	6½	5	1.02 ± 0.49
PRL	0.5 mg/kg	4½	10	1.40 ± 0.26
PRL + $PGF_{2\alpha}$	0.5 mg/kg 0.6 mg/kg	4½	5	0.68 ± 0.26

n: pooled number of rats from 1–6 experiments

Table IV. Effect of indomethacin on PRL binding in the male rat liver

Treatment	n	Specific binding percent per mg protein ± SEM
PRL 0.5 mg/kg	10	11.98 ± 1.54
PRL 0.5 mg/kg + indomethacin 1.3 mg/kg	5	14.10 ± 0.75
PRL 0.3 mg/kg	9	12.21 ± 0.97
PRL 0.3 mg/kg + indomethacin 1.3 mg/kg	5	13.29 ± 1.42
PRL 0.1 mg/kg	5	13.67 ± 1.52
PRL 0.1 mg/kg + indomethacin 1.3 mg/kg	5	15.25 ± 0.53

PGs have been implicated in the mechanism of action of PRL in the mammary gland [for review see 10] and we tested the possibility of their involvement in our system. In table III we see that PGE_1, PGE_2 or $PGF_{2\alpha}$ administered alone, for 6½ days, could not mimic the PRL effect on the induction of its receptor in the male rat liver. Although PGE_2 and $PGF_{2\alpha}$ did cause a doubling of the mean specific binding per mg protein, in comparison with the vehicle injected control group, this increase was not statistically significant. $PGF_{2\alpha}$ was also co-administered with PRL, using a short PRL treatment, in order to evaluate the possibility of PG potentiation of a submaximal PRL effect. Again, no statistically significant effect of $PGF_{2\alpha}$ was observed.

Table IV shows the results of a study of the role of endogenous PGs in the PRL induction of its own receptor in the male rat liver. The maximal non-lethal dose of indomethacin, administered with 3 different doses of PRL for 6½ days, caused no significant effect on PRL binding in the liver, when compared with the respective groups administered PRL only. Doubling the dose of indomethacin to 2.5 mg/kg resulted in the premature death of the rats.

Discussion

The mechanism of action of PRL at its peripheral site of action has been studied in the greatest detail in the mammary glands. *Rillema*'s group [for review see 10] has been largely responsible for the demonstration that PRL stimulation of casein synthesis involves, on the one hand, PRL activation of membranal phospholipase A2 resulting in increased PG synthesis and on the other hand, PRL stimulation along the polyamine synthetic pathway. Both PG and polyamine production are required for PRL stimulation of casein synthesis.

We were interested in the action of PRL in inducing its own receptor in the male rat liver, having previously studied the hormonal regulation of the PRL receptor in the liver in comparison with that in the male sex glands [5]. The male rat, with its low levels of endogenous PRL and female sex hormones, as well as low liver PRL receptor levels, offers a convenient model for the study of the induction of the PRL receptor. The role of PRL in restoring PRL receptors in the liver of hypophysectomized female rats was recently confirmed, using PVP solution to sustain PRL blood levels [15]. Using the same solvent system for the injection of PRL, we have studied the mechanism by which PRL induces its own receptors in the male rat liver. For a given dose level of PRL, the use of 10% PVP in saline as the vehicle for injection was associated with a much greater and reproducible induction of the PRL receptor in the male rat liver, when compared with saline only. These results support the suggestion by *Posner* [16] that the occasional failure of PRL to induce its own binding sites in the liver was largely a failure to produce a chronic and sustained elevation of circulating PRL levels.

The dose-response relationship for the PRL induction of PRL-binding sites in the liver was demonstrated and it was found that even very small doses of PRL (12.5 μg/kg) were effective when administered in 10% PVP. The time relationship of this response was studied: only after 4½ days of twice-daily injections of oPRL was there a significant induction over basal levels, 6½ days being necessary for a high degree of stimulation. Such kinetics for the induction could be seen as reflecting the interdependence between PRL and its own receptor in the liver. At the low initial PRL receptor concentrations, the PRL response is low; as more receptors are induced by PRL, so is the sensitivity and thus the response to it increased. Such an explanation is supported by the much more rapid effect of PRL observed in female rats, where initial PRL receptor levels are high [6]. Further, the hormonal effect being observed is a complex one. The regulation of the PRL receptor in the rat liver has been broadly studied and variously reported to involve some or all of the following hormones: GH, ACTH, thyroid hormones, estrogen, cortisol and testosterone, in addition to PRL itself, which may, in part, be acting in concert with some or all of these to maintain a given level of PRL receptor at a given physiological state [1, 2, 9, 22]. *Scatchard* [21] analysis of the PRL receptor induction as a function of the duration of PRL administration, revealed that the change observed with time is in the PRL receptor concentration and not in the latter's affinity for the hormone. Indeed, there appeared to be a slight increase in receptor concentration already after 2½ days of PRL administration, but this was too low to be statistically significant. Thus, we are not observing an effect of PRL on the conformation of its receptor, but rather a dose- and time-dependent increase in the number of available binding sites.

The importance of endogenous PRL in maintaining the low basal PRL receptor level found in the mature male rat liver was evaluated by administration of bromocryptine or of anti-rPRL antiserum, to lower the effective endogenous hormone levels. These treatments did not cause any further reduction in the already low basal levels of PRL binding, results which are in agreement with those of earlier studies which found no effect of bromocryptine either on the high level of PRL binding in the liver of castrated male rats [2] nor on the low level in the liver of testosterone-treated castrated rats and thus support the notion that testosterone is the dominant modulator maintaining low hepatic receptors for PRL in the male rat [5].

PGs have variously been reported to mediate the peripheral action of PRL. In mammary gland explants, the effect of PRL on uridine incorporation could be blocked by inhibition of PG synthesis by indomethacin and imitated by exogenous addition of $PGF_{2\alpha}$, the PG precursor arachidonic acid [18], or the phospholipid splitting enzyme, phospholipase A2, the maximal effects of which are not additive to those of PRL, suggesting a common mechanism of action [20]. In rat mesenteric blood vessels, PRL is a potent stimulator of PG synthesis and both PRL and PGE_1, or its precursor, dihomogammalinolenic acid, potentiated vascular reactivity at low concentrations and inhibited at high ones [11]. The possibility that PGs may mediate the PRL action in stimulating its own receptor in the liver was therefore tested. At the dose level used, PGE_1, PGE_2 and $PGF_{2\alpha}$, administered in PVP, caused no statistically significant change in the PRL receptor concentration. Furthermore, co-administration of $PGF_{2\alpha}$ and PRL did not increase the rate of induction of PRL-binding sites; this argues against $PGF_{2\alpha}$ being one of several intermediates determining the rate of the effect. The dose of the PGs administered was about half of those found effective in other studies [12, 13]; however, since the PVP increased severalfold the efficacy of indomethacin, a small molecule, as well as that of protein hormones, it is probable that the PG doses used were adequate. These experiments, however, do not exclude the possibility that a PG other than those used in this study is the intermediate of PRL in the liver. In this regard, we examined the involvement of endogenous PGs in this system by co-administration of indomethacin, an inhibitor of the enzyme cyclo-oxygenase, together with PRL. The inhibitor, though given in the maximal non-lethal dose, did not reduce the PRL-induced increase in PRL sites. Since indomethacin eliminates all species of PGs, it may be argued that in the liver, the drug inhibits the formation of two PG types: one that mediates the effect of PRL, and another, a normally present antagonistic PG. This could result in the absence of a net effect of indomethacin. Antagonism between different PG species is found in many systems, including the mammary gland [19]. However, since indomethacin alone did not increase PRL binding in the liver, it is very unlikely that a PG acting to reduce the number of PRL-binding sites plays a physiologically important role in this tissue.

Thus, in this study, PRL was shown to be a potent inducer of its own receptor in the male rat liver and it does not appear to be acting through stimulation of PG synthesis.

References

1 Aragona, C.H.; Bohnet, H.G.; Fang, V.S., and Friesen, H.G.: Induction of lactogenic receptors. II. Studies on the livers of hypophysectomized male rats and rats bearing a growth hormone secreting tumor. Endocr. Res. Commun. *3:* 199–208 (1976).
2 Aragona, C.; Bohnet, H.G., and Friesen, H.G.: Prolactin binding sites in the male rat liver following castration. Endocrinology *99:* 1017–1022 (1976).
3 Barkey, R.J.; Shani, J.; Amit, T., and Barzilai, D.: Specific binding of prolactin to seminal vesicle, prostate and testicular homogenates of immature, mature and aged rats. J. Endocr. *74:* 163–173 (1977).
4 Barkey, R.J.; Shani, J.; Amit, T., and Barzilai, D.: Characterization of the specific binding of prolactin to binding sites in the seminal vesicle of the rat. J. Endocr. *80:* 181–189 (1979).
5 Barkey, R.J.; Shani, J., and Barzilai, D.: Regulation of prolactin binding sites in the seminal vesicle, prostate gland, testis and liver of intact and castrated adult rats: effect of administration of testosterone, 2-bromo-α-ergocryptine and fluphenazine. J. Endocr. *81:* 11–18 (1979).
6 Costlow, M.E.; Buschow, R.A., and McGuire, W.L.: Prolactin stimulation of prolactin receptors in rat liver. Life Sci. *17:* 1457–1466 (1975).
7 del Pozo, E.; Brun del Ré, R.; Varga, L., and Friesen, H.: The inhibition of prolactin secretion in man by CB154 (2-Br-α-ergocryptine). J. clin. Endocr. Metab. *35:* 768–771 (1972).
8 Djiane, J.; Durand, P., and Kelly, P.A.: Evolution of prolactin receptors in rabbit mammary gland throughout pregnancy and lactation. Endocrinology *100:* 1348–1356 (1977).
9 Gelato, M.S.; Marshall, M.; Boudreau, M.; Bruni, J.; Campbell, J.A., and Meites, J.: Effects of thyroid and ovaries on prolactin binding activity in rat liver. Endocrinology *96:* 1292–1296 (1975).
10 Horrobin, D.F.: Prolactin, vol. 5, pp. 59–63 (Churchill/Livingstone, Edinburgh 1977).
11 Karmazyn, M.; Manku, M.S., and Horrobin, D.F.: Changes of vascular reactivity induced by low vasopressin concentrations – interactions with cortisol and lithium and possible involvement of prostaglandins. Endocrinology *102:* 1230–1236 (1978).
12 Lahav, M.; Meidan, R.; Amsterdam, A.; Gebauer, H., and Lindner, H.R.: Intracellular distribution of cathepsin D in rat corpora lutea in relation to reproductive state and the action of prostaglandin $F_{2\alpha}$ and prolactin. J. Endocr. *75:* 317–324 (1977).
13 Lamprecht, S.A.; Herlitz, H.V., and Ahren, K.E.B.: Induction by $PGF_{2\alpha}$ of 20-α-hydroxysteroid dehydrogenase in first generation corpora lutea of the rat. Mol. cell. Endocrinol. *3:* 273–282 (1975).
14 Manku, M.S.; Horrobin, D.F.; Zinner, H.; Karmazyn, M.; Morgan, R.O.; Ally, A.I., and Karmali, R.A.: Dopamine enhances the action of prolactin or rat blood

vessels. Implications for dopamine effect on plasma prolactin. Endocrinology *101:* 1343–1345 (1977).

15 Manni, A.; Chambers, M.J., and Pearson, O.H.: Prolactin induces its own receptors in rat liver. Endocrinology *103:* 2168–2171 (1978).

16 Posner, B.I.: Regulation of lactogen specific binding sites in rat liver: studies on the role of lactogens and estrogen. Endocrinology *99:* 1168–1177 (1976).

17 Posner, B.I.; Kelly, P.A., and Friesen, H.G.: Induction of a lactogenic receptor in rat liver: influence of estrogens and the pituitary. Proc. natn. Acad. Sci. USA *71:* 2407–2410 (1974).

18 Rillema, J.A.: Effects of arachidonic acid on RNA metabolism in mammary gland explants of mice. Prostaglandins *10:* 307–312 (1975).

19 Rillema, J.A.: Effects of prostaglandins on RNA and casein synthesis in mammary gland explants of mice. Endocrinology *99:* 490–495 (1976).

20 Rillema, J.A. and Anderson, L.D.: Phospholipases and the effect of prolactin on uridine incorporation into RNA in mammary gland explants of mice. Biochim. biophys. Acta *428:* 819–824 (1976).

21 Scatchard, G.: The attraction of proteins for small molecules and ions. Ann. N.Y. Acad. Sci. *51:* 660–672 (1949).

22 Sherman, B.M.; Stagner, J.I., and Zamudio, R.: Regulation of lactogenic hormone binding in rat liver by steroid hormones. Endocrinology *100:* 101–107 (1977).

23 Vaitukaitis, J.; Robbins, J.B.; Nieschlag, E., and Ross, G.T.: A method for producing specific antisera with small doses of immunogen. J. clin. Endocr. Metab. *33:* 988–991 (1971).

R.J. Barkey, PhD, Department of Physiological Chemistry and Pharmacology, Roche Institute of Molecular Biology, Nutley, NJ 07110 (USA)

Prolactin-Receptor Dissociation and Down-Regulation

Paul A. Kelly, Jean Djiane and André De Léan

Department of Molecular Endocrinology,
Le Centre Hospitalier de l'Université Laval, Québec; Laboratoire de Physiologie de la Lactation, INRA, Jouy-en-Josas, and Department of Medicine,
Duke University, Medical Center, Durham, N.C.

Specific prolactin (PRL) binding has been identified in crude membrane fraction of many tissues including mammary gland, mammary tumor, liver, kidney, adrenal, ovary, testis, prostate, seminal vesicle, uterus and chorioid plexus of the brain [11]. For a number of hormones (luteinizing hormone, LH, follicle-stimulating hormone, thyrotropin-releasing hormone and LH-releasing hormone), binding to specific receptors located on the plasma membrane is followed by an increase in cyclic AMP production (second messenger) which modulates cellular activity. However, for PRL, insulin and growth hormone (GH), the steps involved in the mechanism of action subsequent to the binding of the hormone to its receptor are yet to be elucidated. For many polypeptide hormones, including some of those which stimulate cyclic AMP production, binding is followed by an internalization of the hormone or hormone-receptor complex [2, 3].

We will presently describe the effect that time of association of labeled PRL with its receptor has on the rate of dissociation of PRL from its receptor. In addition, we will investigate the up- and down-regulation of PRL receptors induced by PRL in both *in vivo* and *in vitro* models.

Measurement of Prolactin Receptors

For the quantification of receptor levels in a tissue, crude plasma membrane fractions are prepared by differential centrifugation, and ovine prolactin is iodinated to a low specific activity (20–60 μCi/μg). PRL binding is assayed using a fixed quantity of membrane preparation with a fixed quantity of [^{125}I]-ovine PRL (oPRL, approximately 100,000 cpm). The difference in

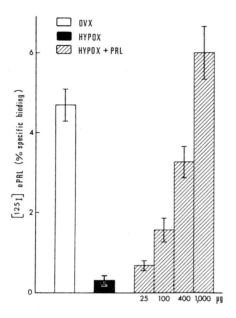

Fig. 1. Effect of hypophysectomy and increasing doses of oPRL (25, 100, 400, and 1,000 μg) in PVP injected twice a day for 7 days to hypophysectomized rats on specific binding of [^{125}I]oPRL to microsomal fraction of rat liver.

the counts per minute of tubes incubated in the absence (total binding) or presence (nonspecific binding) of excess unlabeled PRL yields the PRL specifically bound to the binding sites. Dividing the specific binding by the number of counts per minute added to the incubation results in the 'percent specific binding'. In addition, saturation curves or displacement curves can be carried out and the data transformed into Scatchard plots yielding affinity constants and binding capacities of the membranes [13, 14].

Hormone Regulation of Prolactin Receptors

The hormonal regulation of PRL receptors is complex [11, 12]. Estradiol injection into male or female rats leads to an increase in hepatic PRL-binding sites [15, 21]. The fact that PRL binding can be stimulated by estrogens fluctuates with the estrous cycle and is reduced by ovariectomy implies a direct physiological involvement of estradiol. The loss of PRL binding in rat liver following hypophysectomy implied the importance of a pituitary factor in the maintenance of these binding sites [15, 21].

A direct effect of PRL on its own receptor was first implied when we demonstrated that PRL binding to rat liver in hypophysectomized rats given a pituitary implant under the kidney capsule began to increase approximately 3 days following the increase in serum PRL levels [22]. Figure 1 shows the direct stimulatory effect of PRL (injected in polyvinylpyrrolidone, PVP, to retard adsorption) on PRL-binding sites in rat liver. Twice daily injections of increasing doses of PRL in PVP is capable of completely reversing the hypophysectomy-induced loss of PRL binding, agreeing with observations recently reported [18].

The up-regulatory effect of PRL on PRL receptors in rabbit mammary gland has also been demonstrated [7]. Pseudopregnant rabbits injected with 100 IU oPRL showed a marked increase in PRL receptor levels. This increase could be prevented by simultaneous administration of progesterone, suggesting that part of the progesterone block of lactation during pregnancy could be mediated by a reduction in PRL receptor levels in the mammary gland.

Dissociation of Prolactin from Its Receptor

PRL has a relatively slow rate of dissociation from its receptor. Under conditions in which specific binding has plateaued, 24–48 h are required for 50% dissociation of PRL from its receptor in crude membrane preparations of rat liver [16] and rabbit mammary gland [24].

Effect of Time of Association

The time required for near-maximal association of [^{125}I]ovine PRL with its receptor is approximately 10–12 h. The rate of dissociation of [^{125}I]oPRL from its receptor in rat liver and rabbit mammary gland was studied as a function of time of association. Microsomal fractions were incubated at 23°C in tubes containing a total volume of 0.5 ml for 15 min, 1 or 10 h with $1–3 \times 10^5$ cpm of labeled PRL. Dissociation at 23°C was studied by the addition of 3 ml of buffer containing 10 μg/ml of oPRL. In a separate series of tubes, dissociation was measured from membranes for which association was performed using a saturating concentration (4 nM) of labeled PRL. This was achieved by diluting the [^{125}I]oPRL with unlabeled PRL.

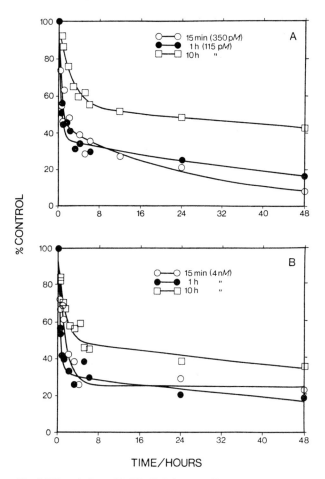

Fig. 2. Dissociation of [^{125}I]oPRL from rat liver membranes. Association was allowed to proceed for 15 min, 1 or 10 h in the presence of 1–3 × 10^5 cpm of labeled PRL. Dissociation was initiated by the addition of 3 ml assay buffer containing 10 μg/ml oPRL. Values are expressed as a percentage of control (time 0) values.

Figure 2a shows the dissociation of labeled PRL from rat liver microsomes using non-saturating concentrations of [^{125}I]oPRL. The dissociation curves can be better fit in a computerized curve fitting model with two kinetic components, the faster component being 60–100 times more rapid than the slower component. For the shorter association times of 15 min and 1 h, 50% dissociation is attained in the first hour, with maximal dissociation of 80–90% at 24–48 h. For an association time of 10 h, dissociation is slower

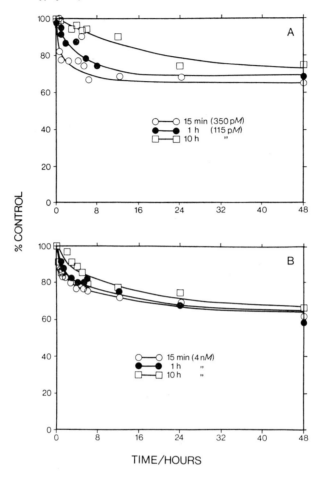

Fig. 3. Dissociation of [^{125}I]oPRL from rabbit mammary membranes. Association and dissociation were performed as described in the legend to figure 2.

due to a 5-fold reduction in the faster kinetic component, with maximal dissociation at 48 h of only 53%. When association is performed in the presence of near-saturating concentrations of PRL (4 nM), a slightly larger percentage of the faster component is observed (fig. 2b).

Figure 3 shows the dissociation of [^{125}I]oPRL from microsomal membrane fractions of rabbit mammary gland incubated with non-saturating (a) or saturating (b) concentrations of labeled PRL. The initial (first component) has a similar rate to that observed in rat liver but, however, the second component is irreversible. Maximal dissociation ranges between 25 and 40%.

In vitro *Desaturation Using* $MgCl_2$

Circulating levels of PRL or placental lactogens are often elevated, exceeding 1,000 ng/ml during pregnancy, lactation or during spontaneous afternoon surges [14]; therefore, a certain fraction of the PRL receptors in target cells might be occupied by endogenous PRL. Although the procedure of homogenization and membrane fractionation offers the opportunity for PRL to dissociate, a large portion of the PRL remains bound to receptors at the end of the preparation, due in large part to the low temperatures used during centrifugation procedures and slow rate of dissociation of PRL from its receptor.

An *in vitro* desaturation technique to dissociate PRL from the receptor was developed [13]. Magnesium chloride at a concentration of 3–5 M is capable of removing 90–95% of the labeled hormone from rabbit mammary gland or rat liver PRL receptors. When the receptors are reexposed to fresh labeled PRL, they retain their ability to specifically bind the PRL [13].

Down-Regulation of Prolactin Receptors

In vivo

In contrast to the inhibitory effect of a large number of hormones on the level of their own receptor [17], a stimulatory effect of PRL on its receptor in both rabbit mammary gland and rat liver has been observed [7, 22]. Using 4 M $MgCl_2$ to dissociate bound PRL from its receptor, we investigated the short-term action of PRL on its receptor in target tissues with the goal of evaluating if PRL, in addition to its ability to up-regulate PRL receptors is, like most other hormones studied so far, capable of inducing a down-regulation of its own receptor. This in turn would lend some support to recent view [23] contending that down-regulation (and possibly up-regulation as well) are ubiquitous events which might be intimately linked to the very mechanism of hormone action.

Lactating New Zealand rabbits were injected every 12 h over a 36-hour period with 2 mg of the dopamine agonist, CB-154, to lower circulating PRL levels [8] after which the animals were anesthetized with 50 mg/kg of sodium pentobarbital. 3 mg bovine PRL (bPRL) was injected intravenously and 2-gram biopsies of mammary gland tissue were removed at the indicated times between 0 and 30 h after PRL injection.

As illustrated in figure 4, injection of 3 mg of PRL leads to a maximal occupancy of free rabbit mammary gland PRL receptors 15 min after the

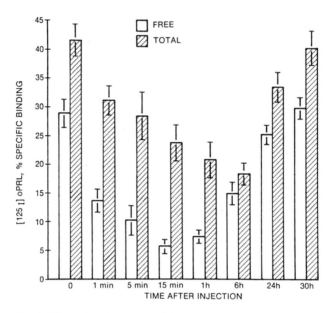

Fig. 4. Effect of an intravenous injection of 3 mg bovine PRL on PRL receptors in rabbit mammary glands. Biopsies (2 g) were removed at the indicated times after PRL injection from lactating rabbits. Free and total (MgCl$_2$-treated) PRL receptor levels were determined. Binding is expressed as a percentage of specific binding per 400 μg protein. Values are means ± SEM of 7 animals.

intravenous injection, corresponding to periods just following maximal serum concentrations. The highest serum levels are seen 1 min after injection with values rapidly declining thereafter. Although saturating concentrations of circulating PRL are present 15 min after injection, 20% of the PRL receptors remain free to bind [^{125}I]oPRL. This could be due to an inaccessibility of the receptors to the circulating PRL or to some dissociation occurring while membranes were isolated from the tissues.

Somewhat surprisingly, total PRL receptor levels assayed following *in vitro* desaturation with 4 M MgCl$_2$ declined progressively up to 6 h after the intravenous injection of PRL and returned to normal at 24–30 h. The difference in total PRL binding between time 0 and 6 h was statistically significant ($p < 0.01$). A difference was observed between the pattern of occupation (free receptors) and the down-regulation reflected by total receptors. In addition, free receptors increased between 1 and 6 h, whereas total receptors continued to decline until 6 h [5]. A similar down-regulation of PRL receptors in rat liver was observed following injection of 1 mg oPRL into ovariectomized, estrogen-treated rats [5].

In vitro

It has been established that the mammary gland can be maintained in organ culture and responds well to hormones. In addition, mammary explants can be used as an experimental model to study the steps involved in the mechanisms of hormone action. PRL receptors are maintained in organ culture for periods up to 48 h. Addition of cycloheximide to the culture media results in a rapid decline of binding. Removal of cycloheximide from the culture medium results in a return of PRL binding to near-control levels 18–24 h later [6].

A down-regulation of PRL receptors in rabbit mammary gland in organ culture has also been observed. Inclusion of PRL (5 μg/ml) in the medium results in a 80% saturation of free receptors. The pattern of the reduction of total PRL receptors is different from that of free receptors with a maximal effect observed in explants cultured in the presence of PRL for 48 h [6].

Effect of Lysosomotropic Agents

As mentioned earlier, for a number of polypeptide hormones, binding is followed by an internalization of the hormone-receptor complex [2, 3], after which the labeled ligands become associated with lysosomal components in the cells [3, 9]. Lysosomotropic agents, such as chloroquine, methylamine or ammonium chloride have been shown to reduce clustering for α_2-macroglobulin and epidermal growth factor (EGF) on cell surface of fibroblasts [19]. [^{125}I]-hCG which has been internalized and is associated with lysosomes is rapidly degraded to mono-iodotyrosine. This process of degradation could be inhibited by lysosomotropic agents [1].

In order to examine if the down-regulation of PRL receptors in rabbit mammary gland involves a lysosomal-mediated step, mammary explants were cultured in the presence of chloroquine (100 μM). Figure 5 shows that PRL binding after 24 h in explant culture in the presence of insulin (In) alone is similar to binding values in explants which were cut but were immediately frozen at -20°C rather than cultured. PRL induces a 53% down-regulation of total PRL receptors. Chloroquine alone markedly increases the basal level of PRL binding and also reduces the ability of PRL to induce

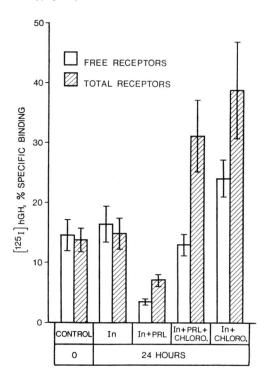

Fig. 5. Effect of PRL and chloroquine (100 μM) on free and total PRL receptor levels in mammary glands in explant culture. Cultures were performed in the presence of insulin (In). PRL binding was measured with [^{125}I]hGH.

a down-regulation of its receptors [4]. A similar effect of ammonium chloride and methylamine reducing PRL-induced down-regulation of PRL receptors is also observed[4].

Conclusions

The events involved in the binding, internalization and cellular action of PRL are summarized in figure 6. Although the initial event of PRL action is the binding to receptors on the plasma membranes, data are accumulating on the localization of PRL and PRL receptors within the target cell [10, 20, 23]. The fact that, with prolonged periods of incubation, microsomal membranes from rabbit mammary gland and rat liver develop a slowly

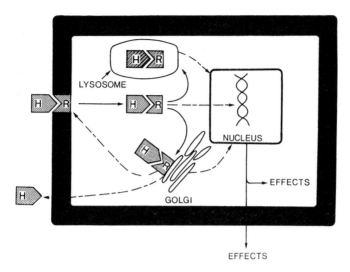

Fig. 6. Summary of possible mechanisms involved in the binding, internalization, and cellular action of a hormone (PRL).

dissociable component implies that these membranes may retain at least some of the capacity of intact cells to internalize hormone and/or hormone receptor complexes, which may represent a heterogeneity or a compartmentalization of PRL receptors occurring subsequent to binding.

The regulation of PRL receptor concentrations in various target tissues is complex. PRL has been clearly shown to have an up-regulatory or stimulatory activity on its own receptor level in both rabbit mammary gland and rat liver [7, 18, 22]. Following the development of a technique to remove endogenous PRL from receptors without destroying receptors, we have been able to demonstrate a transient down-regulation of PRL receptors both *in vivo* and *in vitro*. In addition, using mammary explants, we have shown that PRL receptors appear to have a relatively rapid rate of turnover, based on the rate of receptor replenishment following removal of cycloheximide.

PRL receptors have been localized within the purified Golgi fractions of rat liver. Following injection of [^{125}I]oPRL into female rats, radioactivity was localized in Golgi fractions [10]. The radioactivity eluded from these fractions remained intact, as judged by its ability to bind fresh receptors. The involvement of lysosomes in hormone degradation, which has been clearly demonstrated for other peptide hormones, is also suggested for PRL; therefore, one can postulate that following binding, PRL is internalized after which some of the radioactivity is associated with Golgi elements and some with lysosomes. It is possible that the internalized PRL-receptor complex could act directly at the nuclear level to induce its action. Alternatively, a breakdown product from lysosomal degradation, or some altered form of PRL following association with the Golgi may interact with the nucleus to induce the actions associated with PRL. Hopefully, studies currently in progress should shed some light on the intracellular mechanisms by which PRL regulates its action.

References

1   Ascoli, M. and Puett, D.: Degradation of receptor-bound human chorio-gonadotropin by murine Leydig tumor cells. J. biol. Chem. *253:* 4892–4899 (1978).
2   Bergeron, J.J.M.; Posner, B.I.; Josefsberg, Z., and Sikstrom, R.: The demonstration of specific binding sites for insulin and human growth hormone in Golgi fractions isolated from the liver of female rats. J. biol. Chem. *253:* 4058–4066 (1978).
3   Conn, P.M.; Conti, M.; Harwood, J.P.; Dufau, M.L., and Catt, K.J.: Internalization of gonadotropin-receptor complex in ovarian luteal cells. Nature, Lond. *274:* 598–600 (1978).
4   Djiane, J.; Kelly, P.A., and Houdebine, L.M.: Effects of lysosomotropic agents, cytochalasin B and colchicine on the down-regulation of prolactin receptors in mammary gland explants. Mol. cell. Endocrinol. *18:* 87–98 (1980).
5   Djiane, J.; Clauser, H., and Kelly, P.A.: Rapid down-regulation of prolactin receptors in mammary gland and liver. Biochem. biophys. Res. Commun. *90:* 1371–1378 (1979).
6   Djiane, J.; Delouis, C., and Kelly, P.A.: Prolactin receptors in organ culture of rabbit mammary gland: effect of cycloheximide and prolactin. Proc. Soc. exp. Biol. Med. *162:* 342–345 (1979).
7   Djiane, J. and Durand, P.: Prolactin-progesterone antagonism in self-regulation of prolactin receptors in the mammary gland. Nature, Lond. *266:* 641–643 (1977).
8   Djiane, J.; Durand, P., and Kelly, P.A.: Evolution of prolactin receptors in rabbit mammary gland during pregnancy and lactation. Endocrinology *100:* 1348–1356 (1977).
9   Gordon, P.; Carpentier, J.L.; Cohen, S., and Orci, L.: Epidermal growth factor: morphological demonstration of binding, internalization and lysosomal association in human fibroblasts. Proc. natn. Acad. Sci. USA *75:* 5025–5029 (1978).
10 Josefsberg, Z.; Posner, B.I.; Patel, B., and Bergeron, J.J.M.: The uptake of prolactin into female rat liver: concentration of intact hormone in Golgi apparatus. J. biol. Chem. *254:* 209–214 (1979).
11 Kelly, P.A.; Djiane, J., and De Léan, A.: Interaction of prolactin with its receptor: dissociation and down-regulation; in Scapagnini and MacLeod, Control of peripheral regulation of prolactin function (Raven Press, New York, in press, 1980).
12 Kelly, P.A.; Ferland, L., and Labrie, F.: Endocrine control of prolactin receptors;

in Robyn and Harter, Progress in prolactin physiology and pathology, pp. 59–68 (Elsevier/North-Holland Biomedical Press, Amsterdam 1978).
13 Kelly, P.A.; Leblanc, G., and Djiane, J.: Estimation of total prolactin binding sites after *in vitro* desaturation. Endocrinology *104:* 1631–1638 (1979).
14 Kelly, P.A.; Leblanc, G.; Ferland, L.; Labrie, F., and De Léan, A.: Androgen inhibition of basal and estrogen-stimulated prolactin binding in rat liver. Mol. cell. Endocrinol. *9:* 195–204 (1977).
15 Kelly, P.A.; Posner, B.I., and Friesen, H.G.: Effects of hypophysectomy, ovariectomy and cycloheximide on specific binding sites for lactogenic hormones in rat liver. Endocrinology *97:* 1408–1415 (1975).
16 Kelly, P.A.; Posner, B.I.; Tsushima, T.; Shiu, R.P.C., and Friesen, H.G.: Tissue distribution and ontogeny of growth hormone and prolactin receptors; in Raiti, Advances in human growth hormone research, pp. 567–584 (General Printing Office, Washington 1974).
17 Lesniak, M.A. and Roth, J.: Regulation of receptor concentration by homologous hormone: effect of human growth hormone on its receptor in IM-9 lymphocytes. J. biol. Chem. *251:* 3720–3729 (1976).
18 Manni, A.; Chambers, M.J., and Pearson, O.H.: Prolactin induces its own receptors in rat liver. Endocrinology *103:* 2168–2171 (1978).
19 Maxfield, F.R.; Willingham, M.C.; Davies, P.J.A., and Pastan, I.: Amines inhibit the clustering of a_2-macroglobulin and EGF on the fibroblast cell surface. Nature, Lond. *277:* 661–663 (1979).
20 Nolin, J.M. and Witorsch, R.J.: Detection of endogenous immunoreactive prolactin in rat mammary epithelial cells during lactation. Endocrinology *99:* 949–958 (1976).
21 Posner, B.I.; Kelly, P.A., and Friesen, H.G.: Induction of a lactogenic receptor in rat liver: influence of estrogen and the pituitary. Proc. natn. Acad. Sci. USA *71:* 2407–2410 (1974).
22 Posner, B.I.; Kelly, P.A., and Friesen, H.G.: Prolactin receptors in rat liver: possible induction by prolactin. Science *187:* 57–59 (1975).
23 Posner, B.I.; Raquidan, D.; Josefsberg, A., and Bergeron, J.M.: Different regulation of insulin receptors in intracellular (Golgi) and plasma membranes from livers of obese and lean mice. Proc. natn. Acad. Sci. USA *75:* 3302–3306 (1978).
24 Shiu, R.P.C. and Friesen, H.G.: Properties of a prolactin receptor from the rabbit mammary gland. Biochem. J. *140:* 301–311 (1974).

P.A. Kelly, PhD, Groupe du Conseil de Recherches Médicales en Endocrinologie Moléculaire, Le Centre Hospitalier de l'Université Laval,
2705 boul. Laurier, Québec, G1V 4G2 (Canada)

Discussion

Herington: Your observations on the lack of total reversibility of binding to receptors is quite common with a number of different hormones and their receptors. The explanation of potential internalization is quite nice, but I am not sure that it is totally adequate. We find exactly the same sort of lack of reversibility in totally solubilized receptors; this is particularly with growth hormone, but also for prolactin. This, taken with the fact that you often do not see lack of reversibility in whole-cell preparations, makes me wonder whether internalization is an adequate explanation for the phenomenon.

Kelly: I would agree that internalization may not be the ideal explanation, nor may compartmentalization as I called it. However. *Friesen* had a paper recently showing similar effects, of increasing the time and concentration of prolactin, on the dissociation of prolactin from its receptors, primarily in the rabbit mammary gland. It is interesting to note that in the whole-cell preparation, the isolated hepatic cell for example, your group has also shown that there is a much more rapid rate of dissociation of prolactin from the whole cell than from the membrane preparations. *Shiu*'s data tie in very well with this on mammary cell-lines. Whether there is an actual active processing of prolactin, once it enters the cell, resulting finally in its excretion from the cell is another possibility which should be in investigated.

Waters: We have also looked at solubilized receptors in terms of the dissociation. I do not get the same sorts of effect with growth hormone and insulin dissociation as I get with prolactin. With the prolactin receptor, I get the same phenomenon, even using affinity-purified, that is highly purified, prolactin receptors. I take it from that, that we are dealing with a conformational change. If anyone has any thoughts on how we can look at a conformational change, I should like to hear their ideas.

General Discussion on the Prolactin Target Cell and Receptor
Chairmen: *H. Guyda and I.R. Falconer*

Eastman: Is there any evidence that there are multiple receptor sites, or different classes of receptor sites for prolactin? Or are we seeing a single site, and a beautiful straight-line Scatchard plot?

Kelly: It is quite clear that in a number of experiments you can find a second class of lower-affinity prolactin binding sites, with approximately 50–100 times lower affinity than the higher class. Most people tend to look only for the higher prolactin binding sites in the studies which are published.

Barkey: In our system in the liver, we were quite clearly observing only one order of binding sites.

Eastman: The question really is whether we are deluding ourselves in not looking at the total number of binding sites. I am wondering what is the significance of excluding the lower-affinity binding sites, and looking only at the very high affinity.

Barkey: In one of our earlier studies, we did not separate the two classes. We found that the second class of binding site, the lower-affinity one, was almost parallel to the baseline. In fact, it had such low affinity that it was difficult to work with.

Kelly: I would agree with that. Probably 50% of the time, the lower-affinity site is identical to the nonspecific binding, and you really cannot get a good quantification of it.

Eastman: Are these the sites that are involved in up- and down-regulation? Or are we talking about just one class of sites? One speaker showed us magnificent changes in capacity, they were parallel with the same affinity. Are we seeing just one class coming up all the time, or are there changes and multiple classes?

Kelly: We certainly have data that the changes we see in the number of binding sites are always in the number of high-affinity binding sites. The lower affinity binding sites do not appear to change in number.

Barkey: I would agree with that: we see no change in the number of low-affinity sites whatsoever.

Guyda: What is the physiological role of the lactogenic receptor in the liver? Is bile green milk?

Kelly: There have been a number of reported actions of prolactin in the liver – effects on free fatty acid synthesis, somatomedin production, total RNA synthesis. The critical studies, however, showing the binding of prolactin to its receptors, followed by a specific biochemical event, are yet to be shown for the liver.

Baxter: I wanted to comment on the possible role of the liver prolactin receptor. Perhaps one way of approaching this is to manipulate the receptor concentration, and see what else happens. In our laboratory, we have looked at a series of diabetic rats, with various extents of diabetes. We find a decrease in the liver lactogenic receptor, which has a highly significant correlation with the serum somatomedin level. In fact, we can show that the serum somatomedin level correlates with the hepatic output of somatomedin. This is certainly consistent with there being, although it by no means proves that there is, some relationship. However, we find this only in female rats. In the male, although the somatomedin levels decrease in diabetes, there is a 5-fold increase, or induction, in liver prolactin receptors. If these receptors are involved in generating somatomedin in the female, they certainly do not seem to have a role in the male.

Guyda: What was the nature of the somatomedin assay determination?

Baxter: It was a radioreceptor assay of human placental membrane.

Guyda: Using whole plasma?

Baxter: Yes.

Kelly: Barkey mentioned the use of CB-154 to lower circulating prolactin levels as a means of removing endogenous prolactin, and used a time period of 36 h. It is true that 36 h is the ideal time for the rabbit mammary gland, but it is too long for the rat liver, perhaps because of the difference in the rate of the degradation of prolactin between the two species. If you go longer than 24 h, you start to see a reduction in the level of receptors, probably due to an inhibition of the positive stimulatory effect of prolactin on its receptors. That might explain your reduced effect in the presence of CB-154 treatment.

Barkey: That is a possibility, although we were comparing this with the control animals, which also had 48 h after the last prolactin injection.

Guyda: Last week, Posner proposed that the Golgi is an important source of receptor binding for prolactin. When one looks at receptors on the surface of the plasma membrane, one must realize that there is a number of events taking place in the interior of the cell which influences what one sees on the external surface. It is imperative to understand the system exceedingly well, and to know whether you are dealing with plasma membranes, which have been chopped up and have different microsomal fragments in them, or whether you are, in fact, dealing with whole cell systems, where the interior of the cell might be in a position to regulate that at which you are looking. Receptors can be shed into the external surfaces; receptors can be internalized, and recycled back onto the membrane.

What is a prolactin receptor? Does it have to do something to qualify, or is just the presence of specific binding of a radiolabelled prolactin sufficient?

Greenwood: In our laboratory, we do not use the term receptor until we can link a binding with a biological effect. So, we do not have any receptors, we have a lot of binding proteins.

The other point is that if the receptor is physiological, and we accept down-regulation under chronic stimulation with prolactin, it then has something to say to the morning session as to what it means, biologically, to have high levels of prolactin. The only thing I can think of getting at it is going back to old bioassay data, comparing total amount of hormone injected – whether it is in a single bolus to get a transient level and shock the system because the receptors are there and you do not down-regulate, or whether you give divided doses, with which you can get down-regulation. Perhaps, someone well-versed in prolactin bioassay could actually say whether this is physiological.

Shin: I have a very general question to the audience. When we deal with pharmacologists, they talk about two different theories; one is the rate theory and the other is the occupation theory. Here, we are dealing entirely with the occupation theory, binding versus action; we ignore completely rate theory. I wonder why.

Kelly: Once we see prolactin binding somewhere, it certainly does not mean that it represents a prolactin receptor: we do like to find some action associated with it. But, it does open the possibility that the tissue might be a target for prolactin, and one should certainly try to look for possible actions of prolactin in that tissue.

I would also like to make a comment about the time which prolactin needs to be in contact with its receptor for an action to be seen. Rosen and Houdebine have shown, using the mammary explant system from the rat and the rabbit, that as little as a 4- to 6-hour incubation period of prolactin with the explant system can induce a maximal development of casein, as well as the message for casein, 24 h later. Normally they allow the incubation to proceed for 24 h, but if the prolactin is put in for only 4 h and then removed, there appears to be enough which is, possibly, internalized to induce a subsequent action in terms of production of casein. We are, in turn, now in the process of looking at the number of prolactin receptors which are required to be occupied to induce a certain action in both the rabbit and the rat.

Bern: The thing that concerns me has to do, in the simplest terms, with how fast prolactin can act. The issues come up particularly in studies in which the effects of prolactin on transport are involved. There have been studies on mammalian preparations, and even on our own species, in which prolactin has been said to have immediate renal effects, which are very interesting; that promptly raises another problem, the possible contamination of prolactins with antidiuretic factors, known or unknown, which may be in such preparations.

Quite apart from that, there are papers on the effects of prolactin on transport in vitro, across such structures as the amnion, including the human amnion, the fetal skin, and the fetal urinary bladder. We have done a fair amount of work, over a period of years, on various preparations having to do with the effect of prolactin on transport, in both fish and mammals. In all cases, in the past, where we have taken our preparations, which will respond if the hormone is injected into the animal 1–3 days before we study the preparation, or where organ culture or cell culture techniques are used, the responses occur. But if we take these preparations out of the animal and place them into an incubation system, and then add prolactin, on the basal side or on both sides, and look for an effect, we cannot find any effect. That indicated to us that there were not any immediate actions of prolactin.

Venetia France raised the interesting possibility that what was the matter with our preparations was that they were, essentially, from prolactin-free environments. If we had taken systems which had previously been exposed to prolactin, and had then tested for the effects of additional amounts of prolactin, we might have seen some effects. We took all the model systems we had going in our laboratory. If, for example, we took our mouse mammary epithelium preparation (this is a monolayer culture on collagen gel) and if it has been exposed previously to prolactin, and we put it into an incubation system, we can get no electrophysiological evidence of changed transport. If we take the urinary bladder of a fish, which has been acclimated to 5% seawater (which means that its prolactin has been turned on) we cannot get any indication of an effect of adding prolactin. If we take a piece of skin from the same fish, which extrudes chloride, if we take it from a freshwater fish in which the prolactin has been turned on, again we cannot get any effect of adding prolactin. The posterior intestine of the same fish again showed no response with regard to the addition of prolactin. The same goes for a second species of fish with regard to both the urinary bladder and the opercular membrane. Exposure to prolactin earlier did not allow us to get any reaction to prolactin added to the preparation within 1 to a little over 2 h. Therefore, I raise the question: Are there rapid actions of prolactin, and if there are, why are we not seeing them?

Falconer: I should just like to add some data of ours. The system which we are using, which shows the most rapid responses in vitro to prolactin, is the lactating mammary gland slice from rabbits which have been in mid-lactation treated with bromocriptine for 24–48 h to reduce endogenous prolactin. The tissue is done in a slice incubation system, and we have been measuring, in particular recently, ^{86}Rb transport, which is equivalent to potassium transport. The fastest effect that we can reliably demonstrate is in 20 min, with prolactin added to the incubation medium. We have been looking at it after shorter times, but we cannot pick anything up. With a system of the sensitivity that we are using, 20 min is quite a long time: we were trying to use it to discriminate between the activation of preexisting sodium/potassium pump sites, and the formation of new sites. The time lag would probably indicate that new sites are being synthesized, or new protein is being synthesized which is activating the sites, rather than a direct immediate action of prolactin on the membranes

Rillema: In the cultured mammary gland system, one can see a threefold increase in ornithine decarboxylase activity within 30 min after adding prolactin to that system.

Falconer: One of the things that came out quite clearly is that we are looking at an extremely dynamic system. We have a hormone which is released in a pulsatile manner. The evidence showed very ragged lines for prolactin concentrations with respect to time, whether you did the measurements every hour, by catheter, or every 4 h, or whatever. The reason for the sharply peaked system that is shown is not only that there is pulsatile release, but the turnover of prolactin is extremely fast. Our own data indicate that it turns over with a half-life of about 10 min, and I think that is in agreement with other workers. The clearance rate of prolactin in the blood is very fast indeed, and if you look to see where it has gone, it has not all gone to the mammary gland, or even to the liver; the organ which takes up the bulk of the prolactin is the kidney, and you can find it coming out very quickly in the effluent from the kidney in the ureter. It is degraded into low molecular weight fragments, and the bulk of the prolactin in the body goes through the kidney and comes out as fragments. The organ taking the next biggest share of prolactin out of the system is the liver; and the bulk of the prolactin coming out of the liver, in the bile, is also fragments. It is cleared very quickly: the amount that gets into the mammary gland, even in a lactating animal, is only a few percents of the total which is being secreted. We are looking at a very dynamic system, and we are only looking at a few small parts of what happens in the total picture. The main thing that happens to the prolactin in the circulation is that it gets taken out and degraded and lost. It is only a very small proportion that does anything.

The turnover time of prolactin in liver tissue is much shorter than it is in the mammary gland. This is partly because of the effects that were talked about earlier – that the receptors in the mammary gland, once they have prolactin on them, tend to hold it.

The question of receptor regulation is very interesting, and it is becoming more biologically real. Why tissues respond to hormones in one physiological state and do not respond in another is becoming very apparent in respect to the availability of receptors. We have seen evidence, and data were presented today, of increases in receptors as a consequence of prolactin itself. And we had data previously, in the literature, for increases in prolactin receptors as a result of stimulation by other hormones, particularly the gonadal steroids. We have heard that, in the very short-term, prolactin receptors can be seen to disappear as a consequence of an increase in circulating prolactin.

The question of internalization of hormones is going to be with us for a long time. It seems clear, in a large number of the studies, that the internalization of peptide hormones is a route for their degradation and destruction. Certainly, when one looks to see what is in the cells after hormone has been internalized, one finds low molecular weight fragments of the hormone, in the majority of cases. It could be that some of these are biologically active, and the growth hormone workers have been chasing these biologically active fragments for a long time. I have chased biologically active fragments of prolactin for a couple of years, and have come up with nothing. Nevertheless, there are fragments of prolactin in the cells that take it up. Whether they are of some importance, or are simply on their way out, remains to be demonstrated.

Prolactin can move right through cells. One of the interesting things about milk, particularly at the time of parturition, is that it probably contains appreciably higher prolactin concentrations than there are in the blood. It may have got in between the cells, or it may have got in through the cells, but it has certainly concentrated in the milk. This applies to other hormones too.

The mechanism of action of prolactin is still a mystery. There has been work by Rillema and others on biologically active compounds, which vary in concentration in mammary cells as a consequence of prolactin's action. But to be able to pin down just how they fit together to make a picture which activates both the differentiating effects of prolactin on inducing milk constituents formation in the mammary gland, and how they work on the massive enzymic changes in the mammary gland, which occur as a consequence of prolactin, remains to be seen.

Our own work on ion transport is no clearer. We can demonstrate highly reproducible effects of prolactin on ionic transport mechanisms, and we can block the lactogenic effects of prolactin by using things that interfere with ion transport, such as ouabain and valinomycin, which is an ionophore. It is still very difficult, however, to interprete just how these are working. The point that really stands out is that prolactin is going to provide us all with a livelihood for some years to come!

Prolactin: Questions without Answers

Prolactin: Questions Without Answers

John O. Willoughby[1]

Centre for Neuroscience, Flinders University of South Australia and Flinders Medical Centre, Bedford Park, S.A.

The impetus given to research in prolactin by the recognition of its existence in the human [25] and the development of specific radioimmunoassays in many species, has brought about rapid advances in understanding prolactin regulation, function and pathology. Foremost among advances would be the recognition of the important inhibitory regulation of prolactin by dopaminergic neural mechanisms [34], the identification and understanding of common hyperprolactinaemic states [27], and an appreciation of the various physiologic stimuli which result in its secretion.

In this brief review, attention will be focused upon the rapidly diminishing areas where important aspects of prolactin biology remain incompletely defined, or where there are contentious points of view.

Prolactin: Its Nature

The structure of human prolactin was published by *Shome and Parlow* [61] in 1977. It is a 198 amino acid peptide with a molecular weight of 20,000 daltons. 99 amino acid residues are identical in sequence position in prolactins from human, rat, pig and sheep. Surprisingly, in view of the established prolactin activity of human growth hormone, only 16% of human prolactin amino acid residues have identical sequence positions with human growth hormone and even fewer with human placental lactogen.

What remains an unresolved question is the nature of other substances having prolactin immunoreactivity or bioactivity or both, but which by

---

[1] The author wishes to thank Ms. *V. Frangoulis* for expert secretarial assistance.

various separation techniques appear not to be one of the monomeric prolactin polypeptides referred to above. Essentially, prolactin-like molecules have been separated according to size (chromatography) or electrical charge (electrophoresis) and typified by immuno- or bioactivity. In most studies in which different prolactin-like substances have been observed, the primary interest of investigators has been directed mainly at monomeric immunoreactive prolactin so the nature of the other prolactin-like substances has been relatively neglected.

In most species studied prolactin-like substances in addition to presumed monomeric prolactin have been reported, namely in human [11, 56, 66], rat [2], and monkey [24]. Because the most detailed studies of the relationships between prolactin-like substances have been carried out in the rat, information available from these studies will be viewed in detail. A comprehensive examination of electrophoretically separable forms of rat prolactin has come from the laboratory of Dr. *C.S. Nicoll*, who with his collaborators, utilized a number of bioassays and radioimmunoassays to characterize various forms of prolactin [2]. The results indicated, for example, that purified rat prolactin (NIH RP-1) is comprised of 4 radioimmunoactive, electrophoretically separable, prolactin-like substances of which only one has bioactivity (table 1). Furthermore, five radioimmunoassayable and three bioactive prolactin-like fractions could be obtained by electrophoresis of pituitary tissue or culture medium of pituitary explants.

In those regions where prolactin-like substances were present by both radioimmunoassay and bioassay, the relative quantities differed so that generally much more activity was measured by bioassay than by radioimmunoassay. A number of cautionary conclusions emerge from this work. Firstly, either nonprolactin proteins in NIH RP-1 prolactin cross-react in the radioimmunoassay, or minor structural modifications to prolactin, seriously affecting its bioactivity, occur to account for apparent heterogeneity in this reference preparation. Secondly, there appear to be several forms of prolactin which have different bioassay:radioimmunoassay ratios and that age, sex, reproductive state and *in vitro* culture conditions each result in differences in the prolactin species under study.

In a further study [32] comparing bioassayable and radioimmunoassayable serum prolactin, the common problems of a heterologous bioassay (e.g. *rat* unknown in *pigeon* cropsac using *ovine* standards) were avoided by utilizing rat mammary gland explants in an organ culture as a bioassay system. Triiodothyronine, growth hormone, estrogens and progesterone failed to stimulate secretory activity in this homologous bioassay. It was

Table I. Bioassay (BA) and radioimmunoassay (RIA) estimates of prolactin activity in eluates from different regions of PDGE columns on which purified rat PRL (RP-1) was electrophoresed[1]

Gel segment No.[2]	Prolactin activity, mU/ml[3]		BA/RIA
	BA	RIA	
1	_[4]	_[4]	_[4]
2 (GH)	_[4]	9.2	_[4]
3 (ALB)	_[4]	30.5	_[4]
4 (PRL)	173 (76–335)	128.0	1.3
5	_[4]	27.4	_[4]

[1] 50 μg of RP-1 were applied to each of 10 columns.
[2] Abbreviations in parentheses indicate the principle protein bands present in these segments. ALB = Albumin.
[3] The BA results are presented with the 95% confidence limits in parentheses under the mean potency estimates.
[4] No activity was detected.
Reproduced from *Asawaroengchai et al.* [2] with permission.

demonstrated that, on average, radioimmunoassay detects only about 25% of the prolactin activity measured by bioassay; a remarkably high correlation between bioassay and radioimmunoassay was achieved in selected experiments. Quantitative differences between these assays was again ascribed to heterogeneity of prolactin in serum, and to the fact that radioimmunoassays are developed to measure a highly selected population of prolactin molecules.

Optimistic conclusions drawn by these investigators include that: (a) radioimmunoassay does give measurements of blood prolactin levels that are physiologically meaningful; (b) most (71–93%) immunoassayable prolactin is found in the same electrophoretic band as prolactin standards, and in one experimental model (virgin female's pituitary) the percentage distribution of bioassayable prolactin in different electrophoretic regions was virtually the same as radioimmunoassayable activity. Recent studies in humans agree with the view that prolactin of different forms and receptor binding activities are secreted in different states of pituitary activity [11]. The nature of these radioimmunoassayable prolactin-like substances, however, remains an enigma.

Prolactin Synthesizing Cell Types

It is fully accepted that prolactin in peripheral blood, cerebrospinal fluid and milk is derived from prolactin of pituitary origin. However, considerable quantities of prolactin found *in vivo* have an uncertain site of origin.

A prolactin-like substance, which is in every way identical to prolactin of pituitary origin, is present in amniotic fluid in a concentration 100 times that of plasma [3]. Because maternal hypophysectomy and fetal death do not decrease amniotic fluid prolactin concentrations, attention has focused on amnion, chorion, decidua and trophoblast as potential sources of this prolactin. While it is agreed that amnion and placenta fail to synthesize prolactin in culture, chorionic and/or decidual cells do appear to be capable of sustained prolactin secretion [18]. In fact, *Riddick and Kusmik* [55] regard decidual tissue as the most likely cell of origin of amniotic prolactin, because placenta, amnion, chorion and amniochorion, stripped free of decidual tissue, released only very small amounts of prolactin *in vitro*, compared to decidua. In a recent immunofluorescence study, prolactin was localized in the cytoplasm of especially the decidua cells [12] providing further support for decidua as the cell of origin of amniotic fluid prolactin.

A second situation, of some surprise, is that a prolactin-like substance is present and probably synthesized in the central nervous system. In 1972, *Fuxe et al.* [15] reported localization of prolactin-like immunoreactivity to nerve terminals in rat hypothalamus in several regions, namely, arcuate nucleus, periventricular preoptic area and dorsomedial nucleus. Of further interest was that prolactin-like immunoactivity was even present after hypophysectomy. There is good evidence that growth hormone is synthesized in rat amygdala [47] and ACTH identified in rat brain also appears to be resistant to hypophysectomy [30], thus prolactin may well be yet another systemic peptide to have neuronal localization and function. In further support of this idea is the observation that septal and preoptic neurons are responsive to iontophoretically administered prolactin [50] and locally applied prolactin appears to alter the metabolic activity of dopamine neurons [48].

The last unusual prolactin pool to be considered is the plasma prolactin associated with both prolactin and nonprolactin secreting pituitary tumours. It is well recognized that, in addition to growth hormone, 25% of cases of acromegalics also secrete prolactin and some cases of tumours secreting corticotrophin likewise secrete prolactin. (Mixed hormone-secreting tumours in experimental animals have, of course, been recognized for years.) Furthermore, in 2 patients with hypothyroidism, thyrotropin, prolactin and growth hormone appeared in one tumour [21].

Where careful immunocytology and electron microscopy have been performed, tumours associated with increased levels of more than one pituitary hormone most often contain more than one cell type; for example, prolactin cells and growth hormone cells [21, 23]. Also of interest is one preliminary report that occasional tumours have two hormones within the cytoplasm of one cell [28]. In relation to prolactin secreting tumours, specifically, it is likely that at least some circulating prolactin is derived from nontumorous parts of the pituitary gland. Evidence for this includes morphologic studies in which abundant active prolactin cells occur in the normal gland adjacent to prolactin adenomas [29, 40] and pharmacologic studies in which dopamine antagonists fail to augment prolactin levels in patients with microadenomas. The implication is that 'normal' prolactin cells are already secreting maximally [*Judd*, in preparation]. It has been postulated that large adenomas compress or distort the median eminence or portal vessels such that prolactin inhibitory factors are unable to perfuse the full population of prolactin cells, thus accounting for failure of suppression of nontumour pituitary prolactin secretion. Clearly, such a mechanism does not operate in most instances of small pituitary adenoma and so hyperprolactinaemia from nontumour cells implies instead a fundamental change in prolactin cell regulation. Either tuberoinfundibular neurons in hyperprolactinaemic states are paradoxically inhibited or prolactin cells in a hyperprolactinaemic environment are no longer inhibited by dopamine and related factors. In any case, short loop feedback on nontumorous prolactin cells does not appear to operate! If it is true that nontumorous prolactin cells are maximally secreting prolactin in cases of prolactin adenoma, it is further paradoxic that selective adenomectomy effectively reverses the hyperprolactinaemia.

The Rhythm Section

There are no other hormones with the variety and range of rhythmic phenomena demonstrated by prolactin. For some rhythms the pattern of prolactin secretion can be related to specific environmental or reproductive cycle events, for others there is no identifiable trigger and in any case their function or significance is ill understood.

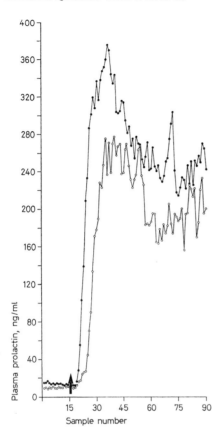

Fig. 1. Prolactin concentrations in blood samples collected every 2 min from rats. The arrow indicates the time of dopamine blockade by pimozide injection. Frequent prolactin bursts become apparent. Reproduced from *Shin and Chi* [60] with permission.

Short Period Rhythms – Less Than 24 h

The prolactin rhythm with the shortest period is revealed best with minute-to-minute blood sampling in animals treated to remove dopamine mediated suppression of prolactin [60] (fig. 1). In such unsuppressed animals, prolactin is secreted in bursts as often as every 8 min. As for other anterior pituitary hormones, a rapidly firing 'pacemaker' generating hormone secretory bursts appears to be present.

The high frequency prolactin secretory pattern referred to above is not seen in the normal prolactin suppressed state. However, in all mammals studied appropriately, less frequent quite distinct pulsatile bursts of secretion are seen, e.g. human [59] and rat [71]. These bursts are adequately measured with 15–20 min blood sampling and in rat (see fig. 2) prolactin bursts may exactly coincide with growth hormone bursts [71]. These bursts of secretion are also generated by neural mechanisms (see next section) and their role in physiology can only be surmised. Perhaps neural signals, because of some fundamental process generating them, are always phasic and metabolic functions through necessity have adapted to briefly periodic pituitary hormone release. Alternatively, episodic hormone secretion may permit pituitary peptides, which compete for similar receptors with other peptides (for example, prolactin and growth hormone), to interact with their receptors during peaks of secretion when competing peptides are not present.

Circadian Rhythms

In the circadian range of frequencies there are four identified prolactin rhythms in the rat, while in other animals some rhythms apparently homologous with those in rat, evidently have different mechanisms generating them.

In rat and man, there is a diurnal variation of prolactin. Previous dispute about the existence of a rat prolactin diurnal rhythm has been resilved: there is a rise in prolactin levels towards the end of the light period which persists through darkness. The rise is probably comprised of an increase in the size and frequency of episodic bursts of prolactin secretion, which in groups of animals is measured as an overall increase in mean prolactin concentration [57, 73]. Whether the rhythm persists in constant light or dark is not established. In man, there is a striking nocturnal increase in prolactin secretion which is closely associated with sleep itself [59] and the mean prolactin rise, again, is a consequence of an increased frequency and amplitude of episodic prolactin bursts. Sleep does not play a role in the nocturnal rise in prolactin in rat (which is nocturnally active) which implies that the two diurnal rhythms are differently generated.

Of special interest in females are those rhythms related to the reproductive cycle which occur with a circadian frequency; there are three. It is well known that there is a surge of prolactin, luteinizing hormone and follicle stimulating hormone on the day of proestrus. This surge can be made to operate *daily* in animals treated with estrogens which simulate the ovarian steroid environment of the proestrous animal [46]. This daily prolactin surge has many of the characteristics of the spontaneous proestrous surge (it is resistant to stress) and indicates that there is a central circadian clock for the surge which is activated by estrogens.

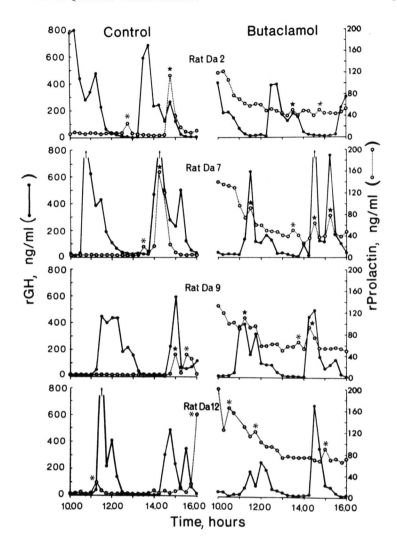

Fig. 2. Concurrent growth hormone (GH) and prolactin concentrations in rats sampled every 15 min for 6 h, before and after dopamine blockade with butaclamol. Infrequent prolactin spikes occur which sometimes coincide with GH spikes. Prolactin peaks are indicated by asterisks and stars; stars indicate concurrence of peak GH and prolactin levels. Reproduced from *Willoughby et al.* [71] with permission.

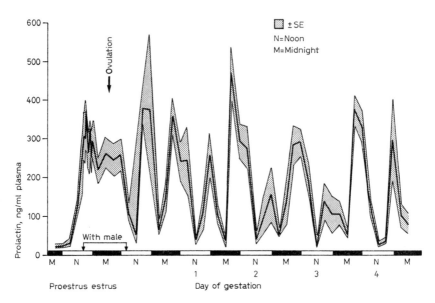

Fig. 3. Plasma prolactin concentrations in groups of rats commencing at proestrous and continuing through early pregnancy. The semicircadian rhythm of prolactin is apparent. Reproduced from *Butcher et al.* [6] with permission.

Different from this artificial proestrous prolactin surge are prolactin rhythms which appear in pregnant or pseudopregnant rats (fig. 3). Two distinct surges of prolactin appear – to which have been applied the term 'semicircadian' – and these surges have been shown to have quite different mechanisms generating them [6, 14, 41]. The daily surge is abolished by stress, while the nocturnal surge is stress-resistant. Neither of these surges conform to the characteristics of the proestrous surge. The stimulus to their initiation does not seem to be cervical stimulation as first thought, and although they can be maintained by progesterone, the function of the corpus luteum (progesterone generation) in turn depends on the semicircadian prolactin surges in early pregnancy [9, 62]. Precisely what neuroendocrine substrate initiates the semicircadian prolactin rhythm is therefore unclear.

Reproductive – Seasonal Rhythms

Prolactin rhythms of periodicity longer than 1 day can be categorized as estrus/menstrual cycle (4–28 days) or as seasonal/circannual rhythms (365 days). The proestrous prolactin surge of the estrus cycle has been extensively studied. It occurs in nearly all mammals and while it is not promi-

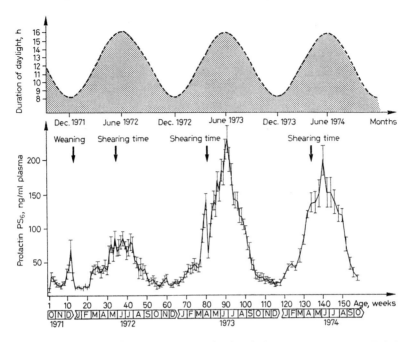

Fig. 4. Seasonal variations in plasma prolactin of Ile de France rams and their relation to day length. Reproduced from *Ravault* [52] with permission.

nent in man, it can be observed [13]. As discussed above, it seems that rising follicular phase estrogens trigger a surge of gonadotrophins and prolactin at proestrous and the latter seems to be an important luteotrophic stimulus in some animals [9, 62]. In humans, at least, there appears to be very little active participation by prolactin in the events of ovulation.

Seasonal prolactin rhythms are specially interesting. They have been best described in ruminants (fig. 4) and in all studies, single morning plasma prolactins vary in proportion to the hours of daylight throughout the year [31, 52]. In addition to actual day length, associated factors such as temperature and humidity appear to participate in altering prolactin concentrations. Increases in *temperature* affect prolactin both through increases in pituitary secretion and by decreases in peripheral clearance [63]. Environmental *humidity* probably acts synergistically with temperature [68]. It is recognized that the pineal gland participates in synchronizing estrus in hibernating animals and that the gland is generally suppressed by light in proportion to the amount of daylight [53]. Because certain day lengths have optimal stimulating effects on plasma prolactin (in comparison to longer or shorter ones) an overall suppressive or facilitatory role for the pineal in seasonal prolactin can be discounted. In partial support of this view is that *Buttle* [7] has found that cervical ganglionectomy (sympathectomy) had only trivial effects on the seasonal prolactin variation in goats (namely, an advance in the occurrence of the spring rise in prolactin).

The existence of these prolactin rhythms is not questioned but their physiologic function is speculative, and the structures generating them largely unknown. Teleological considerations may assist in unravelling their nature.

Prolactin: Pacemakers and Pathways in Brain

There is an extensive literature on neural structures involved in prolactin regulation though studies on structures participating in specific physiologic secretory patterns are infrequent. Some hypotheses regarding brain substrates for prolactin rhythms are now presented.

High Frequency Pulsatile Secretion

It is clear that unsuppressed prolactin is pulsatile [60] with bursts of prolactin occurring every few minutes. All other anterior pituitary hormones are secreted in brief bursts as well and it has been shown in experiments using the Halasz technique [19] that the isolated medial basal hypothalamus is capable of maintaining pituitary hormone secretion and that such secretion is episodic [70, 72]. However, prolactin secretion is suppressed in animals with complete hypothalamic deafferentation [70] in agreement with the idea that dopamine neurons intrinsic to the hypothalamic island continue to act as a prolactin inhibitory system. What is interesting is that pharmacologic blockade of these intrinsic neurons not only augments prolactin secretion, but also induces pulsatile prolactin secretion [*Willoughby and Martin*, unpublished]. Thus, it seems likely that there is a prolactin 'pacemaker' in the hypothalamus, analogous to 'pacemakers' for other anterior pituitary hormones. It would be expected that the tuberoinfundibular cells involved contain prolactin releasing factor, and that local dopamine neurons constitute an inhibitory component (fig. 5).

The fundamental nature of a 'pacemaker' for episodic hormone secretion is not known. Any theory which invokes such a neuronal system ultimately depends on the assumption that neurons, independently or as a network, rhythmically and spontaneously discharge. Important observations relevant to this view are that hypothalamic magnocellular neurosecretory

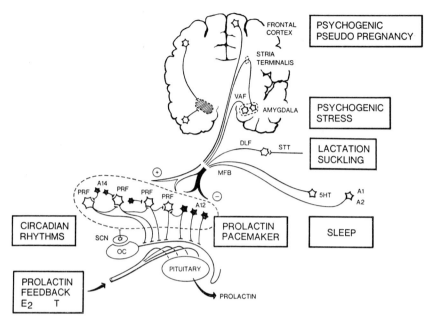

Fig. 5. Neural structures postulated to regulate prolactin secretion. Neural and hormonal inputs influence the activity of a hypothalamic pacemaker for prolactin secretion through actions on its dopaminergic inhibitory or prolactin releasing factor (PRF) components. The physiologic factors which affect prolactin secretion are illustrated adjacent to structures which mediate the effects. DLF = Dorsal longitudinal fasciculus; MFB = medial forebrain bundle; SCN = suprachiasmatic nucleus; STT = spinothalamic tract; VAF = ventral amygdalo fugal pathway; OC = optic chiasm.

neurons have already been shown to discharge in bursts. For example, in response to suckling and dehydration, identified neurohypophyseal neurons discharge bursts of action potentials every few minutes [1, 33]. Unlike magnocellular neurons, the hypothalamic parvicellular tuberoinfundibular neurons do not seem to be situated in focal groups [22]. It would seem implausible, however, that diffusely distributed tuberoinfundibular neurons should fire spontaneously, synchronously and episodically. However, neurons in the ventromedial nucleus of the hypothalamus have been shown by Golgi studies to have a very extensive dendritic network [43] and hypothalamic as well as tuberoinfundibular neurons have been shown by electrophysiological studies to have widely projecting branched axons [20, 54]. This anatomic evidence for extensive intrahypothalamic intercellular communication provides a structural basis for dispersed groups of neurons acting in concert – and possibly providing the fundamental mechanism by which tuberoinfundibular neurons episodically stimulate anterior pituitary hormone secretion.

Episodic Prolactin Secretion

It is unclear what neural mechanisms subserve the commonly observed less frequent spikes of prolactin seen every hour or so in untreated rat and man [72]. Because such spikes do not appear in animals with complete hypothalamic deafferentation an excitatory extrahypothalamic locus must participate in the phenomenon. Both anterior and posterior hypothalamic deafferentations fail to abolish spikes of prolactin secretion, thus the excitatory influence on the 'pacemaker' may enter via the lateral hypothalamus from the medial forebrain bundle. In man, the observation that prolactin spikes are prominent in sleep may suggest that medial forebrain bundle projections from the mesencephalon and raphé ('sleep centres') also participate in spike generation throughout the day. If one uses data from studies utilizing brain stimulation, then most forebrain regions prove to be inhibitory to prolactin secretion – which diminishes the likelihood that such regions participate in prolactin regulation. In further support of the view that the brain stem is the source of neural afferents stimulating prolactin spikes is the observation that pharmacologic agents affecting serotonin and noradrenaline can affect the preovulatory prolactin surge [65] (see fig. 5).

Surges of Proestrus and of Pregnancy and Pseudopregnancy

The preovulatory prolactin surge which operates daily in estrogen-treated animals, has been more extensively investigated. It is clear that a region anterior to the hypothalamus is responsible for triggering the proestrous surge because surgical cuts between the hypothalamus and preoptic area completely abolish the surges [69]. In fact, electrical stimulation and lesion studies suggest that the medial preoptic area is an important structure initiating the proestrous surge. However, rats with medial preoptic area lesions, which spare the anterior hypothalamic area, exhibit estrous cycles separated by pseudopregnancies [8] indicating that participation of the medial preoptic area is not obligatory in ovulation.

In a study of the semicircadian prolactin surges of pseudopregnancy, *Freeman et al.* [14] determined, using animals with knife cuts caudal to the suprachiasmatic nucleus, that an area rostral to the knife cut was important for at least the nocturnal surge. Although not specifically examined for in the study by *Clemens et al.* [8] the fact that pseudopregnancies were maintained in animals with medial preoptic area lesions suggests that the anterior hypothalamic area itself, which was spared by their lesions, is a sufficient neural substrate for both the proestrous prolactin surge and the nocturnal semicircadian prolactin surge of pseudopregnancy.

Because the proestrous surge and the surges associated with pregnancy and pseudopregnancy have a definite relationship to time of day and appear to occur on many consecutive days, it is clear that visual information has an influence on the rhythms. Because of the defined role of the suprachiasmatic nucleus in other circadian rhythms, it is likely that this nucleus regulates the timing of prolactin surges as well. It is now known that these rhythms persist in a continuous light (not dark) environment as shown by *Bethea and Neill* [75] and therefore they would have an endogenous generator governing them (see fig. 5).

Finally, pharmacologic studies point to the participation of mesencephalic serotonin and pontomedullary noradrenaline neurons in the proestrous surge because agonists or antagonists of these transmitters, respectively, enhance or inhibit the proestrous prolactin peak [65].

Diurnal Rhythm
An increase in the magnitude and frequency of episodic bursts of prolactin secretion at night probably accounts for nycthemeral variations in prolactin in normal humans and rats and if pontomesencephalic sleep centres contribute to such a variation, then clearly the influence in man is opposite from the rodent. Suprachiasmatic nucleus involvement is likely.

Seasonal Prolactin
The seasonal rhythms of prolactin are less well investigated. An apparent link between ambient temperature changes and prolactin regulation through the multifunctional medial preoptic area (important in thermoregulation) suggests itself, but no evidence is available. Perhaps the combination of altered temperature and day length acts synergistically to dictate changes in mean prolactin levels via actions respectively in the medial preoptic area and through the suprachiasmatic nucleus on the anterior hypothalamic area. There is already some evidence that longer photoperiods and higher temperatures may be synergistic [49].

Lactation
Tindal and Knaggs [67] have extensively examined pathways involved in prolactin secretion in conscious, decorticate lactating rats. They were able to map two pathways by determining brain sites which responded to electrical

stimulation by evoking prolactin secretion. Both pathways terminated in the medial preoptic-anterior hypothalamic area. The pathway most likely to subserve prolactin release due to suckling is thought to ascend in the spinothalamic sensory projection, then via the region of the dorsal longitudinal fasciculus and medial forebrain bundle, finally reaching the lateral preoptic area from whence it enters the medial preoptic area (see fig. 5). The other pathway defined by *Tindal and Knaggs* [67] is described below.

Stress

It is well known that psychological stresses can trigger prolactin secretion. Many clinicians regard pseudopregnancy as a psychologic condition, and elevated prolactin levels are a feature of this syndrome. Both situations clearly indicate an influence of higher thought processes on hypothalamic function but there are only a few well-defined pathways which potentially could provide a neural substrate for such an influence. Essentially, these are the projections of limbic system structures (amygdala, hippocampus and septum) to the hypothalamus [51]. The amygdala has heavy projections to the hypothalamus via the stria terminalis and ventral amygdalofugal pathway [51]. *Gloor* [17] has postulated that the amygdala assigns to environmental stimuli 'affective significance' for the organism. Such significance then determines the response of the organism to the stimuli – flight or approach, etc. – and it can be presumed that neuroendocrine responses would be integrated with behavioural responses, probably via the stria terminalis and ventral amygdalofugal pathway. Accordingly, the amygdala might mediate neuroendocrine responses to higher thought processes such as psychologic stress and even the intense desire to be or fear of being pregnant. The view that the cerebral cortex can influence endocrine function is further supported by the observations of *Tindal and Knaggs* [67], that frontal cortex stimulation triggers prolactin secretion in lactating rats, and the finding of *Ohta and Oomura* [46] that there is a monosynaptic pathway from frontal cortex to hypothalamus (see fig. 5).

Prolactin: Teleology

There is a large literature supporting the view that prolactin is important in the adaptation of fish to freshwater – specifically by promoting sodium retention and reducing water permeability in membranes exposed to water such as the gills, skin, bladder, gut and kidney (see *Bern* [4]) (fig. 6).

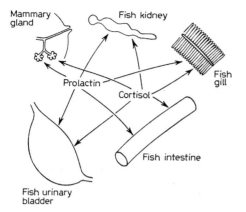

Fig. 6. Osmoregulatory role of prolactin (and cortisol) in fish and mammals. Reproduced from *Bern* [5] with permission.

Stimulated by these observations a number of investigations have shown that prolactin influences a bewildering array of physiological functions related to salt and water balance, blood pressure and cardiovascular control. *Mainoya* [35] has demonstrated an increase in intestinal absorption of fluid and electrolytes in response to prolactin. Early studies also suggested a role for prolactin in regulating osmolarity in subprimate mammals; however, subsequent reports indicate that consistent changes in prolactin secretion in response to enforced changes in plasma osmolarity (or vice versa) do not occur [39]. The role of prolactin receptors, present in high concentration in the kidney remains unclear, even though changes in renal prolactin receptor levels may occur in response to dehydration [38].

Indirectly related to salt and water homeostasis is the maintenance of cardiac output and blood pressure. Evidence from *Horrobin* and co-workers has implicated prolactin directly in magnifying the contractile responses of rat aorta smooth muscle to adrenaline [37] and in stimulating myocardial contractility and heart rate in rabbit [44] although this latter effect was completely dependent on seasonal and sexual factors. Subsequently, dopamine was shown to potentiate both the excitatory effects of low prolactin concentrations and the inhibitory effects of high prolactin concentrations on rat smooth muscle contractions [36]. Finally, *Falconer and Rowe* [10] have shown that prolactin acts to increase salt secretion in breast. There is some evidence, therefore, that prolactin does participate in extracellular fluid space regulation in species other than fish and particularly has an influence on the electrolyte composition of breast milk.

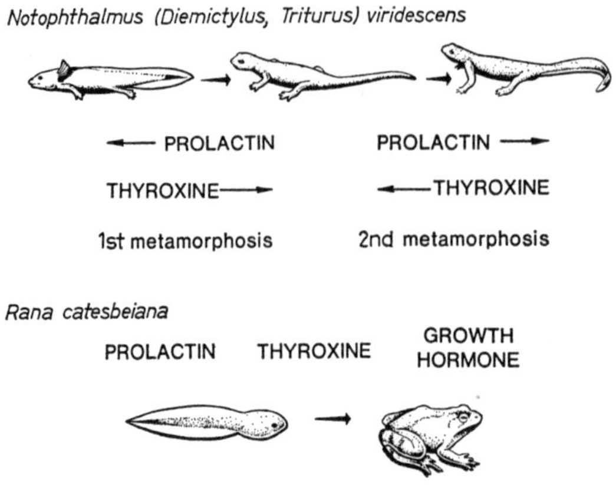

Notophthalmus (Diemictylus, Triturus) viridescens

```
        ← PROLACTIN          PROLACTIN →
        THYROXINE →          ← THYROXINE
        1st metamorphosis    2nd metamorphosis
```

Rana catesbeiana

```
    PROLACTIN   THYROXINE   GROWTH
                            HORMONE
```

Fig. 7. Effects of prolactin on metamorphosis in the newt and frog. Reproduced from *Bern* [4] with permission.

In studies designed to reveal an action of prolactin in human osmoregulation, results have been equivocal. *Rubin et al.* [58] suggest that changes in renal function seen during sleep (for example) might be related to the strong sympathetic activation that accompanies REM sleep episodes, rather than to prolactin secretory bursts or ADH, or aldosterone. *Sowers et al.* [64] examined the effects of osmolarity changes on prolactin secretion and found a synergistic effect of an oral waterload on the prolactin response to TRH and little else of note.

In contrast to the above, there is good evidence that prolactin in amniotic fluid does influence the environment of the developing fetus. *Josimovich et al.* [26] showed in rhesus, that amniotic-fluid prolactin acted to produce net fluid and sodium transfer from the fetus to the maternal side of the amnion. A further action of amniotic fluid prolactin may be in growth and development of the fetus, an idea stimulated by the fascinating effects of prolactin in conjunction with other hormones on development in other classes (see *Meier* [42] and *Bern* [4]) (fig. 7). What is especially interesting is that effects on development depend on phasic prolactin variations, so that in experimental studies, prolactin has contrary effects when injected at different times of the day!

In Salamander, prolactin can initiate a drive to return to the aquatic environment with accompanying structural changes and when injected at certain times of the day it is more effective in producing this effect than at other times.

In frogs, development from the tadpole can be inhibited by prolactin provided that prolactin is administered at an appropriate time of day. Thus,

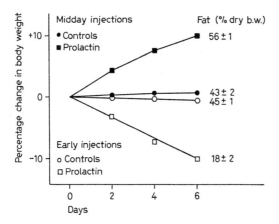

Fig. 8. Changes in body weight and fattening responses in the golden topminnow after injection of prolactin at different times of the photoperiod. Reproduced from *Meier* [42] with permission.

changes in the phase of the rhythm of prolactin release appear to determine whether prolactin inhibits metamorphosis or not [42].

In fish, prolactin injections result in fat deposition or mobilization depending on the time of day prolactin is administered (fig. 8) or the time of injection in relation to other administered hormones such as thyroxin or cortisol.

In birds, prolactin may have either antigonadal effects or may augment gonadal steroidogenesis according to the time of day that prolactin is administered. Even seasonal effects on reproductive activity can be stimulated by exogenous prolactin and corticosterone injections at the appropriate times of day. Furthermore, in some birds, the patterns of endogenous prolactin and corticosterone rhythms are consistent with the conclusion that changing relationships between the two hormonal rhythms control reproduction, metabolism and migratory behaviour.

These observations might point to the importance of prolactin in fetal growth. What is clear is that prolactin rhythms are prominent in lower phyla and appear crucial in regulating development. It is of interest that the semi-circadian prolactin surges of pregnancy in rat resemble prolactin variations in some fish. Perhaps endogenous and exogenous prolactin rhythms function entirely to co-ordinate all events relating to reproductive cycling, fetal growth, nurturing, maturation and associated behaviours. Certainly, the existence of so many prolactin rhythms is intriguing.

References

1. Arnauld, E.; Dufy, B., and Vincent, J.D.: Hypothalamic supraoptic neurons: rates and patterns of action potential firing during water deprivation in the unanaesthetised monkey. Brain Res. *100:* 315–325 (1975).
2. Asawaroengchai, H.; Russell, S.M., and Nicoll, C.S.: Electrophoretically separable forms of rat prolactin with different bioassay and radioimmunoassay activities. Endocrinology *102:* 407–414 (1978).
3. Ben-David, M. and Chrambach, A.: Preparation of bio- and immunoactive human prolactin in milligram amounts from amniotic fluid in 60% yield. Endocrinology *101:* 250–261 (1977).
4. Bern, H.A.: On two possible primary activities of prolactins: osmoregulatory and developmental. Verh. dt. zool. Ges. *1975:* 40–46 (1975).
5. Bern, H.A.: Prolactin and osmoregulation. Am. Zool. *15:* 937–948 (1975).
6. Butcher, R.L.; Fugo, N.W., and Collins, W.E.: Semicircadian rhythm in plasma levels of prolactin during early gestation in the rat. Endocrinology *90:* 1125–1127 (1972).
7. Buttle, H.L.: The effect of anterior cervical ganglionectomy on the seasonal variation in prolactin concentration in goats. Neuroendocrinology *23:* 121–128 (1977).
8. Clemens, J.A.; Smalstig, E.G., and Sawyer, B.D.: Studies on the role of the preoptic area in the control of reproductive function in the rat. Endocrinology *99:* 728–735 (1976).
9. De Grief, W.J. and Zeilmaker, G.H.: Regulation of prolactin secretion during the luteal phase in the rat. Endocrinology *102:* 1190–1197 (1978).
10. Falconer, I.R. and Rowe, J.M.: Effect of prolactin on sodium and potassium concentrations in mammary alveolar tissue. Endocrinology *101:* 181–186 (1977).
11. Farkouh, N.H.; Packer, M.G., and Frantz, A.G.: Large molecular size prolactin with reduced receptor activity in human serum: high proportion in basal state and reduction after thyrotropin-releasing hormone. J. clin. Endocr. Metab. *48:* 1026–1032 (1979).
12. Frame, L.T.; Wiley, L., and Rogol, A.D.: Indirect immunofluorescent localization of prolactin to the cytoplasm of decidua and trophoblast cells in human placental membranes at term. J. clin. Endocr. Metab. *49:* 435–437 (1979).
13. Franchimont, P.; Dourcy, C.; Legros, J.J., et al.: Prolactin levels during the menstrual cycle. Clin. Endocrinol. *5:* 643–650 (1976).
14. Freeman, M.E.; Smith, M.S.; Nazian, S.J., and Neill, J.D.: Ovarian and hypothalamic control of the daily surges of prolactin secretion during pseudopregnancy in the rat. Endocrinology *94:* 875–882 (1974).
15. Fuxe, K.; Hokfelt, T.; Eneroth, P.; Gustafsson, J., and Skett, P.: Prolactin-like immunoreactivity: localization in nerve terminals of rat hypothalamus. Science *196:* 899–900 (1977).
16. Gallo, R.V.; Johnson, J.H.; Goldman, B.D.; Whitmoyer, D.I., and Sawyer, C.H.: Effects of electrochemical stimulation of the ventral hippocampus on hypothalamic electrical activity and pituitary gonadotropin secretion in female rats. Endocrinology *89:* 704–713 (1971).
17. Gloor, P.: Inputs and outputs of the amygdala: What the amygdala is trying to tell the rest of the brain; in Livingston and Hornykiewics, Limbic Mechanisms: the

continuing evolution of the limbic system concept, pp. 189–210 (Plenum Press, New York 1978).

18 Golander, A.; Hurley, T.; Barrett, J.; Hizi, A., and Handwerger, S.: Prolactin synthesis by human chorion-decidual tissue: a possible source of prolactin in the amniotic fluid. Science 202: 311–313 (1978).
19 Halasz, B. and Pupp, L.: Hormone secretion of the anterior pituitary gland after physical interruption of all nervous pathways to the hypophysiotropic area. Endocrinology 77: 553–562 (1965).
20 Harris, M.C. and Sanghera, M.: Projection of medial basal hypothalamic neurons to the preoptic anterior hypothalamic areas and the paraventricular nucleus in the rat. Brain Res. 81: 401–411 (1974).
21 Heitz, P.U.: Multihormonal pituitary adenomas. Hormone Res. 10: 1–13 (1979).
22 Hokfelt, T., et al.: Aminergic and peptidergic pathways in the nervous system with special reference to the hypothalamus; in Reichlin, Baldessarini and Martin, The hypothalamus, pp. 69–135 (Raven Press, New York 1978).
23 Horvath, E.; Kovacs, K.; Ryan, N.; Singer, W., and Ezrin, C.: Incidence of prolactin-producing adenomas. IRCS Med. Sci. 5: 447 (1977).
24 Hummel, B.C.W.; Brown, G.M.; Hwang, P., and Friesen, H.G.: Human and monkey prolactin and growth hormone: separation of polymorphic forms by isoelectric focusing. Endocrinology 97: 855–867 (1975).
25 Hwang, P.; Guyda, H., and Friesen, H.: A radioimmunoassay for human prolactin. Proc. natn. Acad. Sci. USA 68: 1902–1906 (1971).
26 Josimovich, J.B.; Merisko, K., and Boccella, L.: Amniotic prolactin control over amniotic and fetal extracellular fluid water and electrolytes in the rhesus monkey. Endocrinology 100: 564–570 (1977).
27 Kleinberg, D.L.; Noel, G.L., and Frantz, A.G.: Galactorrhea: a study of 235 cases, including 48 with pituitary tumours. New Engl. J. Med. 296: 589–600 (1977).
28 Kovacs, K.; Horvath, E., and Kerenyi, N.A.: Prolactin cell adenomas associated with the multiple endocrine adenomatosis syndrome. IRCS Med. Sci. 5: 446 (1977).
29 Kovacs, K.; Ryan, N.; Horvath, E.; Ezrin, C., and Penz, G.: Abundance of prolactin cells in non-tumorous parts of human pituitaries harbouring prolactin cell adenomas. IRCS Med. Sci. 5: 37 (1977).
30 Krieger, D.T.; Liotta, A., and Brownstein, M.J.: Presence of corticotropin in brain of normal and hypophysectomised rats. Proc. natn. Acad. Sci. USA 74: 648–652 (1977).
31 Leining, K.B.; Bourne, R.A., and Tucker, H.A.: Prolactin response to duration and wavelength of light in prepubertal bulls. Endocrinology 104: 289–294 (1979).
32 Leung, F.C.; Russell, S.M., and Nicoll, C.S.: Relationship between bioassay and radioimmunoassay estimates of prolactin in rat serum. Endocrinology 103: 1619–1628 (1978).
33 Lincoln, D.W. and Wakerley, J.B.: Factors governing the periodic activation of supraoptic and paraventricular neurosecretory cells during suckling in the rat. J. Physiol. 250: 443–461 (1975).
34 MacLeod, R.M.: Regulation of prolactin secretion; in Martini and Ganong, Frontiers in neuroendocrinology, vol. 4, pp. 169–194 (Raven Press, New York 1976).
35 Mainoya, J.R.: Effects of bovine growth hormone, human placental lactogen and ovine prolactin on intestinal fluid and ion transport in the rat. Endocrinology 96: 1197–1202 (1975).

36 Manku, M.S.; Horrobin, D.F.; Zinner, H.; Karmazyn, M.; Morgan, R.O.; Ally, A.I., and Karmali, R.A.: Dopamine enhances the action of prolactin on rat blood vessels. Implications for dopamine effects on plasma prolactin. Endocrinology *101:* 1343–1345 (1977).

37 Manku, M.S.; Nassar, B.A., and Horrobin, D.F.: Effects of prolactin on the responses of rat aortic and arteriolar smooth muscle preparations to noradrenaline and angiotensin. Lancet *ii:* 991–994 (1973).

38 Marshall, S.; Gelato, M., and Meites, J.: Serum prolactin levels and prolactin binding activity in adrenals and kidneys of male rats after dehydration, salt loading, and unilateral nephrectomy. Proc. Soc. exp. Biol. Med. *149:* 185–188 (1975).

39 Mattheij, J.A.M.: Evidence that prolactin plays no role in osmoregulation and sodium balance in the rat. J. Endocr. *72:* 34 (1977).

40 McKeel, D.W. and Jacobs, L.S.: Non-adenomatous pituitary mammotroph hyperplasia in patients with pathologic hyperprolactinemia. Proc. Soc. Endocr., Chicago, A 136 (1977).

41 McLean, B.K. and Nikitovitch-Winer, M.B.: Cholinergic control of the nocturnal prolactin surge in the pseudopregnant rat. Endocrinology *97:* 763–770 (1975).

42 Meier, A.H.: Chronophysiology of prolactin in the lower vertebrates. Am. Zool. *15:* 905–916 (1975).

43 Millhouse, O.E.: The organisation of the ventromedial hypothalamic nucleus. Brain Res. *55:* 89–105 (1973).

44 Nassar, B.A.; Horrobin, D.F.; Tynan, M.; Manku, M.S., and Davies, P.A.: Seasonal and sexual variations in the responsiveness of rabbit hearts to prolactin. Endocrinology *97:* 1008–1013 (1975).

45 Neill, J.D.: Comparison of plasma prolactin levels in cannulated and decapitated rats. Endocrinology *90:* 568–572 (1972).

46 Ohta, M. and Oomura, Y.: Inhibitory pathway from the frontal cortex to the hypothalamic ventro-medial nucleus in the rat. Brain Res. Bull. *4:* 231–238 (1979).

47 Pacold, S.T.; Kirsteins, L.; Hojvat, S.; Lawrence, A.M., and Hagen, T.C.: Biologically active pituitary hormones in the rat brain amygdaloid nucleus. Science *199:* 804–806 (1978).

48 Perkins, N.A. and Westfall, T.C.: The effect of prolactin on dopamine release from rat striatum and medial basal hypothalamus. Neuroscience *3:* 59–63 (1978).

49 Peters, R.R. and Tucker, A.H.: Prolactin and growth hormone responses to photoperiod in heifers. Endocrinology *103:* 229–234 (1978).

50 Poulain, P. and Carette, B.: Actions of iontophoretically applied prolactin on septal and preoptic neurons in the guinea pig. Brain Res. *116:* 172–176 (1976).

51 Raisman, G. and Field, P.M.: Anatomic considerations relevant to the interpretation of neuroendocrine experiments; in Martini and Ganong, Frontiers in neuroendocrinology, pp. 1–44 (Oxford University Press, New York 1971).

52 Ravault, J.P.: Prolactin in the ram: seasonal variations in the concentration of blood plasma from birth until three years old. Acta endocr., Copenh. *83:* 720–725 (1976).

53 Reiter, R.J.: Pineal and associated neuroendocrine rhythms. Psychoneuroendocrinology *1:* 255–263 (1976).

54 Renaud, L.P.: Tubero-infundibular neurons in the hypothalamus of the rat: electrophysiological evidence for axon collaterals to hypothalamus and extrahypothalamic areas. Brain Res. *105:* 59–72 (1976).

55 Riddick, D.H. and Kusmik, W.F.: Decidua: a possible source of amniotic fluid prolactin. Am. J. Obstet. Gynaec. *127:* 187–190 (1977).

56 Rogol, A.D. and Rosen, S.W.: Alteration of human and bovine prolactin by a chloramine-T radioiodination: comparison with lactoperoxidase iodinated prolactins. J. clin. Endocr. Metab. *39:* 379–382 (1974).

57 Ronnekleiv, O.; Krulich, L., and McCann, S.M.: The effect of pinealectomy on the pattern of prolactin secretion in conscious freely moving male rats. Neuroendocrinology *27:* 279–290 (1978).

58 Rubin, R.T.; Poland, R.E.; Gouin, P.R., and Tower, B.B.: Secretion of hormones influencing water and electrolyte balance (antidiuretic hormone, aldosterone, prolactin) during sleep in normal adult men. Psychosom. Med. *40:* 44–59 (1978).

59 Sassin, J.F.; Frantz, A.G.; Weitzman, E.D., and Kapen, S.: Human prolactin: 24 hour pattern with increased release during sleep. Science *177:* 1205–1207 (1972).

60 Shin, S.H. and Chi, H.J.: Unsuppressed prolactin secretion in the male rat is pulsatile. Neuroendocrinology *28:* 73–81 (1979).

61 Shome, B. and Parlow, A.F.: Human pituitary prolactin: the entire linear amino-acid sequence. J. clin. Endocr. Metab. *45:* 1112–1115 (1977).

62 Smith, M.S.; McLean, B.K., and Neill, J.D.: Prolactin: the initial luteotropic stimulus of pseudopregnancy in the rat. Endocrinology *98:* 1370–1377 (1976).

63 Smith, V.G.; Hacker, R.R., and Brown, R.G.: Effect of alterations in ambient temperature on serum prolactin concentrations in steers. J. Anim. Sci. *44:* 645–649 (1977).

64 Sowers, J.R.; Hershman, J.M.; Skowsky, W.R.; Carlson, H.E., and Park, J.: Osmotic control of the release of prolactin and thyrotropin in euthyroid subjects and patients with pituitary tumors. Metabolism *26:* 187–192 (1977).

65 Subramanian, M.G. and Gala, R.R.: The influence of cholinergic, adrenergic and serotonergic drugs on the afternoon surge of prolactin in ovariectomized estrogen treated rats. Endocrinology *98:* 842–848 (1976).

66 Suh, H.K. and Frantz, A.: Size heterogeneity of human prolactin in plasma and pituitary extracts. J. clin. Endocr. Metab. *39:* 928–935 (1974).

67 Tindal, J.S. and Knaggs, G.S.: Pathways in the forebrain of the rat concerned with the release of prolactin. Brain Res. *119:* 211–221 (1977).

68 Tucker, H.A. and Wettemann, R.P.: Effects of ambient temperature and relative humidity on serum prolactin and growth hormone in heifers. Proc. Soc. exp. Biol. Med. *151:* 623–626 (1976).

69 Weiner, R.J.; Blake, C.A., and Sawyer, C.H.: Integrated levels of plasma LH and prolactin following hypothalamic deafferentation in the rat. Neuroendocrinology *10:* 349–357 (1972).

70 Willoughby, J.O.; Terry, L.C.; Brazeau, P., and Martin, J.B.: Pulsatile growth hormone, prolactin and thyrotropin secretion in rats with hypothalamic deafferentation. Brain Res. *127:* 137–152 (1977).

71 Willoughby, J.O.; Brazeau, P., and Martin, J.B.: Pulsatile growth hormone and prolactin: effects of (+) butaclamol, a dopamine receptor blocking agent. Endocrinology *101:* 1298–1303 (1977).

72 Willoughby, J.O. and Martin, J.B.: The role of the limbic system in neuroendocrine regulation; in Livingston and Hornykiewics, Limbic mechanisms: the continuing evolution of the limbic system concept, pp. 211–261 (Plenum Press, New York 1977).

73 Willoughby, J.O.: Pinealectomy mildly disturbs prolactin and growth hormone secretory rhythms in rat. J. Endocr. (in press, 1980).
74 Yamada, Y.: Effects of iontophoretically applied prolactin on unit activity of the rat brain. Neuroendocrinology *18:* 263–271 (1975).
75 Bethea, C.L. and Neill, J.D.: Prolactin secretion after cervical stimulation of rats maintained in constant dark or constant light. Endocrinology *104:* 870–876 (1979).

Dr. J.O. Willoughby, Centre for Neuroscience, Flinders University of South Australia, Bedford Park, S.A. 5042 (Australia)

Discussion

Thorner: I was fascinated by the connection which you demonstrated responsible for the rhythms in the rat and the rise in prolactin. Do you know whether, if you cut those connections, you still get a suckling-induced rise in prolactin?

Willoughby: It is possible to interrupt the suckling-induced rise in prolactin concentration by lesions and cuts in the lateral hypothalamus, adjacent to the medial preoptic area.

Bern: If there are these kinds of circadian and ultradian cycles, the effects ought to be visualizable in such areas as transport, for example. I do not think anyone has ever looked into whether a membrane works differently in the morning from in the evening. When it comes to such aspects as intestinal transport during sleep as opposed to during waking hours, this may be of some real physiological meaning, at least potentially. I would like to raise that as a possible area for investigation. If prolactin is going up and down, and prolactin is doing something about inducing its own receptors, or other receptors, this too ought to be reflected, if these are relatively short-term events. I do not know of work in that area, except that there is a certain amount of study on the differential sensitivity of target tissues to hormones, in organ culture, when they have been removed in the morning as opposed to, say, around midnight. The differences in sensitivity appear to be real. To say that they are inherent in the tissue does not say very much; if one has any belief in receptor concepts, this may be much more meaningful.

Neill: In our studies, what we have shown, of course, is that cervical stimulation, and we now use electrical stimulation that lasts for only 10 sec, can induce this rhythm which can last for as long as 18 days. We have linked this to a memory process. That upsets psychologists, but what is happening is that, after stimulation, the information seems to be stored in the brain, and then it is recalled periodically – actually twice per day – in a rhythm probably set by the superchiasmatic nucleus. We have lesioned the suprachiasmatic nucleus, which totally destroys these surges, they immediately disappear. So, it is clear that the suprachiasmatic nuclei stimulate these surges, as well as determining the time at which they appear. In another suty, we have lesioned the medial preoptic area, in animals whose cervixes have not been stimulated, and the surges bein: this suggests that the medial preoptic area normally restrains the secretion of these surges, and seems to be the area in which we should look for the storage of the 'memory'.

Willoughby: It is clear that the double-daily surges of prolactin can be induced by a non-somatosensory means, and that prolactin administration, by way of a subcapsular extrapituitary graft, will induce the appearance of these rhythms. There are also other ways, which do not involve neural stimulation of any part of the brain. It seems to me, therefore, that the rhythm is already there; it does not require a 'memory' system, and once initiated it will continue until hormonal factors, or other, dictate that it degenerates again.

The other point that I should like to raise is the significance of the development of pseudopregnancy in rats with medial preoptic area lesions, when the lesions are tiny and would only affect around 10% of the area. With such tiny lesions, which are made using a technique with iron or steel electrodes, I would feel, personally, that there are more complex things going on than just a simple preoptic area lesion.

Buttle: With regard to the seasonal change which you see in ruminants, it is an underlying rhythm that occurs in both males and females, and the changes involved in pregnancy and lactation are apparently superimposed on top of the seasonal rhythm. The only point that you did not mention is that the ruminants undergo the seasonal change without any apparent effects of this afferent action of prolactin on a target tissue, and that, at the moment, some people are incriminating the pineal as having some regulatory role on the seasonal change in prolactin rhythm. Later, a considerable time after pinealectomy, the seasonal responses to the changes in day length become disorientated.

Willoughby: I actually read your own paper on that subject, and I was singularly unimpressed by anything that the pineal might do to these rhythms. If I recall your data correctly, pinealectomy advanced the occurrence of this equally dramatic rhythm by a very short period.

Buttle: I only examined the goats for 6 months; one has to examine the animals for considerably longer than this, 2 or more years.

Lyons: What about the luteotrophic substance? It seems to me that *Neill*'s graph showed that there was a surge of lactogenic hormone at the time the corpus formed in the rat and, of course, it persisted through pseudopregnancy. I have always contended that that was the luteotrophic hormone in the rat.

Neill: We have studied those surges very carefully to see whether they were luteotrophic. We did the studies by inhibiting the surges with bromergocriptine and the corpora lutea do regress. If we administer ovine prolactin with the bromergocriptine, we can negate the effect of bromergocriptine. We have shown very clearly that these surges are the luteotrophic hormone, that luteinizing hormone is not involved, and that the old story, which was built by people like *Lyons*, remains true. Prolactin is the luteotrophic hormone in the rat.

Lyons: The fact of the matter is that it is also the luteolytic hormone. If you allow the animal to go a couple of days without its pituitary, this same hormone, which we call luteotrophic, is luteolytic. It is a paradoxical situation, but it is the same chemical substance that, when the animal has lost the use of the pituitary for 2 or 3 days, is luteolytic. We established it not only for the pituitary luteotrophic or lactogenic hormone, but also for the placental lactogen.

Neill: I think that more recent work suggests that prolactin is luteolytic only in the sense of causing regression of a structure which has ceased to secrete progesterone. The old definition was something which inhibits the secretion of progesterone, and I do not think that prolactin qualifies as a luteolytic agent in that sense.

Lyons: After 2 days, it is the continuous injection of that hormone which does cause the complete disappearance of the luteal body.

Prolactin and Transport in Fishes and Mammals

Howard A. Bern, Christopher A. Loretz and Chester A. Bisbee

Department of Zoology and Cancer Research Laboratory,
University of California, Berkeley, Calif.

Research on transport in our laboratory is concerned with the delineation of possible common features of the action of prolactin on various transporting epithelia. In teleost fishes, where an osmoregulatory role of prolactin was first discovered, transporting organs may be influenced by prolactin in several ways (table I). Although the urinary bladders of the flounders *Kareius bicoloratus, Platichthys stellatus* and *P. flesus* show increased sodium and/or chloride absorption from their lumens as a result of prolactin administration, the bladder of the flounder *Paralichthys olivaceus* does not so respond [17–19]. A physiological role for prolactin may be indicated in some species inasmuch as sodium and chloride absorption is increased in *Kareius* [18] and *P. stellatus* [7] following freshwater adaptation, a process favored by endogenous prolactin [1]. Using electrophysiological, radiotracer and gravimetric techniques, we have shown that prolactin is without effect on ion transport across the urinary bladder of the goby *Gillichthys mirabilis* [9, 21]. Prolactin replacement in hypophysectomized *Gillichthys* produces no alteration in the short-circuit current (representing net sodium transport in excess of chloride transport from lumen to serosa). The action of prolactin on the urinary bladder of *Paralichthys* and *Gillichthys* may be entirely to reduce permeability to water, a universal action in euryhaline species thus far examined [9, 17, 18]. The jaw skin of *Gillichthys* [22] and (sub)opercular membrane of *Sarotherodon mossambicus* [16] and *Fundulus heteroclitus* [6] actively extrude chloride ion; using techniques similar to those described above, prolactin appears to inhibit this extrusion [16, 22, 23].

The results from fish studies suggested that prolactin might play a role in ion transport across mammary epithelium which is engaged in producing milk, a fluid whose sodium content appears to be closely regulated [20].

Table I. Effects of prolactin on teleost urinary bladder and skin transport systems

	Urinary bladder			Skin and opercular membrane Cl extrusion	References
	ATPase	NaCl absorption	water permeability		
Flounders (seawater)					
Kareius bicoloratus	0	↑	↓		18, 27
Platichthys stellatus	↑	↑	↓		18, 19, 27
Platichthys flesus		↑	↓		17
Paralichthys olivaceus	0		↓		18
Goby (seawater)					
Gillichthys mirabilis	0		↓	↓	9, 21, 22
Killifish (seawater)					
Fundulus heteroclitus				↓	23
Tilapia (seawater)					
Sarotherodon mossambicus				↓	16

Monolayers of prelactating and lactating mouse mammary epithelium maintained on floating collagen gels exhibit extensive ultrastructural differentiation [5, 10, 12] and secrete casein or lactose, respectively, in response to prolactin, the physiologically appropriate hormone [5, 11]. Biochemically and ultrastructurally differentiated cell cultures seemed an ideal system in which to investigate the possibility that prolactin modulates mammary transepithelial ion transport as suggested by prior experiments on mammary tissue [14, 20]. Figure 1 shows previously reported electrophysiological and tracer flux data obtained from prelactating cultures [4]. Within 1 day, prolactin supplementation of culture medium causes a statistically significant increase in transepithelial potential difference (basal side ground) and short-circuit current and causes a statistically significant decrease in transepithelial resistance. These electrophysiological changes can be induced by medium concentrations of prolactin as low as 1 ng/ml [*Bisbee*, unpublished]. In insulin plus cortisol cultures, tracer fluxes show that net active sodium absorption occurs and is equal to the short-circuit current. Prolactin exposure results in a 3.5-fold increase in sodium absorption, but only a doubling of short-circuit current. There is no statistically significant net chloride movement in cultures from either hormonal treatment group. Therefore, prolactin must induce active transport of other, as yet unidentified, ion(s). Similarly, prolactin significantly increases short-circuit current in cultures of lactating [1.5-fold; 2] and tumorous [2.3-fold; 3] epithelium,

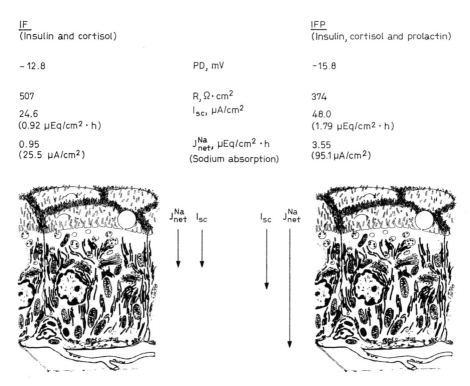

Fig. 1. Summary of the influence of prolactin on cultured mouse mammary epithelium [4]. Medium concentrations of hormones are: insulin, 10 μg/ml; cortisol, 5 μg/ml; prolactin, 10 μg/ml. All prolactin-induced changes are statistically significant differences. There is no statistically significant active chloride movement in cultures from either group. Arrows indicate direction and relative values of net electrical charge or sodium movement across the cultured epithelium. PD = Transepithelial potential difference; R = transepithelial resistance; I_{sc} = short-circuit current; J^{Na}_{net} = net sodium flux (in this case, sodium absorption). PD is measured relative to the basal or blood side of the epithelium.

but does not affect short-circuit current in resting (nonpregnant, nonlactating) cell cultures [3]. When the above electrophysiological results are supported by data showing ouabain- and amiloride-induced reductions of short-circuit currents in prelactating, lactating, and tumor cell cultures [3], we suggest that active sodium transport is present and modulated by prolactin at least in these three physiological states of the gland. The increased sodium absorption caused by prolactin may be due to elevated Na^+K^+-ATPase levels, as has been postulated for mammary glands [1, 13, 15] and as is the case in *Platichthys stellatus* urinary bladders [27]. However, it must be noted that increased transepithelial sodium absorption is not necessarily tightly coupled to increases in Na^+K^+-ATPase levels [26].

The effect of prolactin at various osmoregulatory surfaces is not solely on water permeability and on active sodium transport as initially suggested [1, 14]. Prolactin apparently regulates mineral metabolism in most vertebrates, but the mechanisms affected vary widely from species to species. In searching for possible transport-regulating effects of prolactin in vertebrates, the possibilities of effects on chloride transport [16, 21, 22], on coupled (electrically silent) NaCl transport [7], on divalent ion movements [Ca^{++}; 24, 25], and on transport of other, unidentified ions, or even of no effect on ion movements, all need to be considered.

Acknowledgments

Aided by research grants NSF PCM-78-10348, NIH CA-09041 and CA-05388, and by a NIH postdoctoral fellowship AM-05918 to C.A.L. We are indebted to Dr. *Terry Machen* for his continuing advice.

References

1 Bern, H.A.: Prolactin and osmoregulation. Am. Zool. *15:* 937–948 (1975).
2 Bisbee, C.A.: Active ion transport in cultured mammary epithelial cell monolayers. Fed. Proc. *38:* 1056 (1979).
3 Bisbee, C.A.: Prolactin effects on ion transport across mammary epithelium *in vitro;* PhD thesis, University of California, Berkeley (1979).
4 Bisbee, C.A.; Machen, T.E., and Bern, H.A.: Mouse mammary epithelial cells on floating collagen gels: transepithelial ion transport and effects of prolactin. Proc. natn. Acad. Sci. USA *76:* 536–540 (1979).
5 Burwen, S.J. and Pitelka, D.R.: Secretory function of lactating mouse mammary epithelial cells cultures on collagen gels. Expl Cell Res. *126:* 249–262 (1980).
6 Degnan, K.J.; Karnaky, K.J., and Zadunaisky, J.A.: Active chloride transport in the *in vitro* opercular skin of a teleost (*Fundulus heteroclitus*), a gill-like epithelium rich in chloride cells. J. Physiol., Lond. *271:* 155–191 (1977).
7 Demarest, J.R. and Machen, T.E.: Active Na and Cl transport by the euryhaline teleost urinary bladder (UB). Fed. Proc. *38:* 1059 (1979).
8 Dharmamba, M. and Maetz, J.: Branchial sodium exchange in seawater-adapted *Tilapia mossambica:* effects of prolactin and hypophysectomy. J. Endocr. *70:* 293–299 (1976).
9 Doneen, B.A.: Water and ion movements in the urinary bladder of the gobiid teleost

Gillichthys mirabilis in response to prolactins and to cortisol. Gen. compar. Endocr. *28:* 33–41 (1976).

10 Emerman, J.T.; Burwen, S.J., and Pitelka, D.R.: Substrate properties influencing ultrastructural differentiation of mammary epithelial cells in culture. Tissue Cell *11:* 109–119 (1979).

11 Emerman, J.T.; Enami, J.; Pitelka, D.R., and Nandi, S.: Hormonal effects on intracellular and secreted casein in cultures of mouse mammary epithelial cells on floating collagen membranes. Proc. natn. Acad. Sci. USA *74:* 4466–4470 (1977).

12 Emerman, J.T. and Pitelka, D.R.: Maintenance and induction of morphological differentiation in dissociated mammary epithelium on floating collagen membranes. In Vitro *13:* 316–328 (1977).

13 Falconer, I.R.; Forsyth, I.A.; Wilson, B.M., and Dils, R.: Inhibition by low concentrations of ouabain of prolactin-induced lactogenesis in rabbit mammary-gland explants. Biochem. J. *172:* 509–516 (1978).

14 Falconer, I.R. and Rowe, J.M.: Possible mechanism for action of prolactin on mammary cell sodium transport. Nature, Lond. *256:* 327–328 (1975).

15 Falconer, I.R. and Rowe, J.M.: Effect of prolactin on sodium and potassium concentrations in mammary alveolar tissue. Endocrinology *101:* 181–186 (1977).

16 Foskett, J.K.; Bern, H.A., and Machen, T.E.: Prolactin inhibits chloride secretion by opercular membranes from seawater (sw)-acclimated tilapia *sarotherodon mossambicus*. Fed. Proc. *39:* 492 (1980).

17 Hirano, T.: Effects of prolactin on osmotic and diffusion permeability of the urinary bladder of the flounder, *Platichthys flesus*. Gen. compar. Endocr. *27:* 88–94 (1975).

18 Hirano, T.; Johnson, D.W.; Bern, H.A., and Utida, S.: Studies on water and ion movements in the isolated urinary bladder of selected freshwater, marine and euryhaline teleosts. Comp. Biochem. Physiol. *45A:* 529–540 (1973).

19 Johnson, D.W.; Hirano, T.; Bern, H.A., and Conte, F.P.: Hormonal control of water and sodium movements in the urinary bladder of the starry flounder, *Platichthys stellatus*. Gen. compar. Endocr. *19:* 115–128 (1972).

20 Linzell, J.L.; Peaker, M., and Taylor, J.C.: The effects of prolactin and oxytocin on milk secretion and on the permeability of the mammary epithelium in the rabbit. J. Physiol., Lond. *253:* 547–563 (1975).

21 Loretz, C.A.: Initiation by cortisol of fish urinary bladder electrogenic sodium transport. Amer. Zool. *20:* (in press 1980).

22 Marshall, W.S. and Bern, H.A.: Ion transport across the isolated skin of the teleost, *Gillichthys mirabilis*; in Lahlou, Epithelial transport in the lower vertebrates, pp. 337–350 (Cambridge University Press, London 1980).

23 Mayer-Gostan, N. and Zadunaisky, J.: Inhibition of chloride secretion by prolactin in the isolated opercular epithelium of *Fundulus heteroclitus*. Bull. Mt Desert Is. Biol. Lab. *18:* 106–107 (1978).

24 Neville, M.C. and Peaker, M.: The secretion of calcium and phosphorus into milk. J. Physiol., Lond. *290:* 59–62 (1979).

25 Pang, P.K.T.; Schreibman, M.P.; Balbontin, F., and Pang, R.K.: Prolactin and pituitary control of calcium regulation in the killifish, *Fundulus heteroclitus*. Gen. compar. Endocr. *36:* 306–316 (1978).

26 Schultz, S.G.: Is a coupled Na-K exchange 'pump' involved in active transepithelial

sodium transport? A status report; in Hoffman, Membrane transport processes, vol. 1, pp. 213–227 (Raven Press, New York 1978).
27 Utida, S.; Kamiya, M.; Johnson, D.W., and Bern, H.A.: Effects of freshwater adaptation and of prolactin on sodium-potassium-activated adenosine triphosphatase activity in the urinary bladder of two founder species. J. Endocr. *62:* 11–14 (1974).

Prof. H.A. Bern, Department of Zoology, University of California, Berkeley, CA 94720 (USA)

Discussion

Falconer: In our studies using mammary gland slices, we could only find a chloride shift when the cells had been preloaded with chloride by keeping them in the cold. There were no consistent chloride shifts induced by prolactin, other than that, but it was fairly dramatic when it occurred. In the case of the mammary gland, the milk is a predominantly potassium-rich fluid, rather than a sodium-rich fluid, and the function of the ionic pumps in the mammary gland appears to be to generate enough intracellular potassium to provide the milk with potassium. Sodium seems to be simply extruded from the mammary gland, much as it was extruded from some of your tissues.

Greenwood: Can I ask the obvious question? If you believe in receptors, and measure them, would this not be a nice model system by which to equate the magnitude of the events with the presence or absence of these mythical beings which other people are measuring.

Bern: We have made our own, very serious efforts, even to the point of using the, unfortunately, small amount of fish prolactin that we can get from *Farmer* and *Papkoff* to look for possible receptors, for example in the urinary bladder. We could not find any. That does not mean that they are not there; it has something to do with precision of methods, and so forth. Efforts to find sheep prolactin receptors on gills, for example, where they really should be, have also been fruitless.

Prolactin Secretion Patterns in Patients with Mammary Tumors. Preliminary Results[1]

H.G. Bohnet, A.K. Gabel and P. Kreutzer

Department of Obstetrics and Gynecology, Westfalian Wilhelms-University, Münster

There is clear evidence that the growth rate of experimental tumors of the rat mammary gland can be either stimulated or diminished by alterations of serum prolactin (PRL) [19]. In women with breast cancer no consistent pattern of PRL secretion has been observed [1, 9, 12, 20, 21]. Elevated PRL levels were reported to occur in relatives of patients with breast cancer [15, 17]. In most of these studies, however, serum PRL concentrations were estimated in a single blood sample only. Recently we have shown [6–8] that PRL stimulation tests using metoclopramide (MTCL) will give better information on PRL secretion than basal serum concentrations, since peak PRL levels induced by MTCL correlate with physiological increases of PRL, such as sleep-dependent PRL elevations [7]. This study was designed to investigate PRL secretion in patients with either fibrocystic disease of the mammary gland or with breast cancer.

Material and Methods

Patients with palpable mammary tumors, who were admitted from several outpatient clinics, were subjected to a PRL stimulation test. The time of admission was chosen by the patient. A short time interval (no longer than 6 weeks) between diagnosis and biopsy of the mammary tumor was, however, strongly recommended by the physician. In all patients a mammography had been carried out at least 3 months prior to admission;

---

[1] The studies were supported by the German Research Foundation (DFG, grant No. Bo 511/3). We thank Dr. H. Nienhaus, Institute of Pathology, for the pathohistological investigations and gradings. We are indebted to Dr. F.K. Beller and Dr. H. Wagner for including their patients in this investigation.

the result of this had been openly discussed with the patient, even if a cancer was suspected and mastectomy had been taken into consideration as surgical treatment. Only individuals entered the study who had not received any medication for at least 3 months. In the early afternoon of the day of admission, which in general was the day before surgery took place, the test was performed. A cannula was inserted into an antecubital vein and immediately thereafter (–10) as well as 10 min later (time 0) a blood sample was taken; then a bolus injection of 10 mg MTCL (Paspertin®, Kali-Chemie, Hannover, Germany) was given and further blood samples were withdrawn 25 and 45 min after MTCL administration. In some of the patients the test was repeated 4–6 weeks after surgery, before any further treatment started. Serving as premenopausal controls, ovulating women underwent the same test procedures either during the midfollicular or the midluteal phase of the menstrual cycle; ovulation was judged from a biphasic basal body temperature as well as from serum progesterone (P) levels exceeding 35 nmol/l. These controls consisted out of medical staff and of volunteering patients admitted for tubal sterilization. Postmenopausal controls were obtained by volunteers, who underwent hysterectomy because of incontinence and/or myomas of the uterus; nurses also volunteered for the study. Menopause was considered to have occurred, if in women aged at least 48 years, menstruation had not been observed for at least 1 year and FSH levels exceeded 1,000 μg/l (LER-907) and serum estradiol (E_2) was less than 100 pmol/l [2].

Serum PRL levels were determined in all samples withdrawn using a homologous radioimmunoassay as described elsewhere [3, 11]; results were expressed as mU/l of the MRC standard 71/222. Serum E_2 as well as P was measured at time 0 using a radioimmunoassay; methods applied as well as serum levels observed during the menstrual cycle are given elsewhere [2, 3, 6, 13]; concentrations are given in pmol/l and nmol/l, respectively.

The data obtained were grouped according to the reproductive state of the women, e.g. pre- and postmenopausal. The premenopausal subjects were further subdivided in those tested during the follicular phase and in those during the luteal phase (beyond day 14) of the menstrual cycle.

The pathohistological investigations of the mammary tissue removed by surgery were carried out by one pathologist only. The criteria applied for classification of the histological findings are the following: mastopathy I – elementary dystrophy (not associated with any specific anatomic lesions or any epithelial hyperplasia); mastopathy II – organized dystrophy with regular proliferation of tubular or acinous glandular elements (adenosis) and connective tissue sclerosis; mastopathy III – organized dystrophy but atypical proliferation and connective tissue sclerosis. Statistical analyses were performed using Student's t-test and/or the Kruskal-Wallis rank test.

Results

The serum PRL levels observed in controls and in patients with mammary tumors are given in table I. In premenopausal controls (aged 21–37 years), stimulated PRL levels were significantly ($p < 0.05$, Student's t-test) higher during the luteal than during the follicular phase of the menstrual

Table I. Basal (−10, 0) as well as MTCL-stimulated (+25, +45) serum PRL levels (mean ± SD) in premenopausal (during the follicular and luteal phase of the menstrual cycle) and postmenopausal women with mammary tumors. Either normally ovulating or postmenopausal women served as controls. In premenopausal individuals tested during the luteal phase progesterone (P) levels are also given. Groups were further arranged according to the histological findings of the mammary tissue removed: FA = fibroadenoma; MP I–III = mastopathy; Ca = carcinoma

	Follicular phase (PRL mU/l)					Luteal phase (PRL mU/l)				P nmol/l		Postmenopausal women (PRL mU/l)				
	−10	0	+25	+45		−10	0	+25	+45			−10	0	+25	+45	
Controls	314	309	6,118	5,033	n=11	357	332	8,203	6,859	47.1	n=20	218	193	4,140	3,659	n=18
(SD)	69	64	956	920		83	63	1,413	1,144	11.3		45	46	1,178	1,256	
FA	491	455	6,083	5,489	n=16	565	526	6,735	5,084	25.3	n=16	651	505	6,561	5,503	n=6
(SD)	435	401	2,521	2,323		227	219	2,505	2,071	9.5		400	329	2,396	2,143	
MP I	359	326	7,063	6,325	n=7	631	627	6,903	6,255	20.1	n=9	354	318	5,750	5,046	n=3
(SD)	95	109	1,604	1,345		225	185	2,386	2,027	18.1		95	93	1,008	705	
MP II	433	433	8,609	7,377	n=9	481	438	8,247	7,121	32.3	n=18	545	558	10,662	8,526	n=3
(SD)	170	109	2,234	2,393		261	306	2,897	2,240	12.4		292	152	1,875	2,431	
MP III	347	295	8,084	6,639	n=3	1,080	897	9,812	7,810	16.3	n=12	222	245	7,033	5,878	n=3
(SD)	168	107	856	767		1,011	843	4,056	2,457	6.4		112	119	5,087	3,877	
Ca	386	373	6,810	6,693	n=8	1,415	1,378	8,510	6,942	7.3	n=16	522	467	5,935	5,465	n=20
(SD)	142	140	3,102	2,933		1,224	1,353	3,405	2,749	5.2		454	406	1,451	1,103	

cycle. A trend towards basal PRL elevations was seen during the luteal phase. When 3 of the volunteers were retested in the follicular as well as in the luteal phase of the cycle following the first test cycle, no significant differences of PRL, E_2, and P were observed (data not shown). The age of the premenopausal patients ranged from 19 to 51 years; in the subjects tested during the follicular phase, serum PRL tended to be lower than in those tested during the luteal phase of the cycle. Women with MP III or breast cancer (Ca) exhibited significant luteal elevations of basal PRL ($p < 0.05$, Student's t-test) when compared to controls ($p < 0.001$, Kruskal-Wallis test).

Luteal P concentrations observed in patients with mammary tumors were lower than those seen during the midluteal phase of normally ovulating women. This was clearly pronounced in patients with MP III or cancer (MP III, Ca: $p < 0.05$, Student's t-test; Ca: $p < 0.001$, Kruskal-Wallis test).

In postmenopausal controls aged 49–59 years basal as well as MTCL-stimulated PRL levels were lower than in premenopausal individuals ($p < 0.05$, Student's t-test). After menopause the patients (aged 49–74 years) with benign mammary tumors exhibited slight elevations of PRL when compared to the controls. In women with breast cancer, basal ($p < 0.05$) as well as peak PRL levels seen after MTCL stimulation were significantly elevated ($p < 0.01$, Student's t-test; both basal and stimulated PRL: $p < 0.01$, Kruskal-Wallis test). In subjects with breast cancer (n = 4), who were retested immediately before chemotherapy was started, only slight changes of PRL secretion occurred (data not shown).

Discussion

The results obtained in this study indicate that in patients with mammary tumors serum PRL levels are increased to different extents. This confirms observations of other investigators [9, 12, 15, 17, 20, 26]. The underlying mechanisms, however, are not clear. In premenopausal women with mammary disease, however, there seems to exist a preponderance of endogenous estrogen activity, since luteal P secretion is inadequate; this may result in elevations of PRL as proposed by others [11, 25; see also 7]. Psychic as well as physical stress has been reported to result in short-lived elevations of PRL; permanently increased serum PRL levels due to such stimuli are, however, not known [for references see 7]. In this study serum PRL concentrations were somewhat increased immediately after insertion of the indwelling catheter (at time −10) and decreased with 10 min thereafter (time 0). Thus, it appears that the serum PRL levels observed in the patients with mammary tumors are rather permanently than transiently elevated. In addition, retesting of some individuals did not reveal major changes of PRL secretion which is in agreement with other investigators. This is underlined by observations in infertile women, who underwent similar test procedures and treatment trials of inadequate P secretion; no significant changes of serum PRL occurred [8]. An imbalance between estrogen and P secretion was previously reportes to exist in women with either benign and malignant breast disease [14, 22–24]. Such changes in the hormonal profile are also well known to be a result of increased serum PRL and to cause female infertility [for references see 3, 6, 7]. This could also explain the observations of *MacMahon et al.* [18] that patients with breast cancer do not conceive very easily. The elevations of serum PRL seen in postmenopausal women remain obscure and need further clarification.

In animal experiments estrogens together with lactogenic hormones have been shown to induce PRL as well as estrogen receptor sites in several target organs of these hormones [4, 5, 10]. Thus, it appears that in premenopausal women PRL may exert its effect on two levels: firstly, it inhibits normal luteal formation and secondly it stimulates cell proliferation by induction of estrogen and PRL receptors in the mammary gland.

The preliminary results gained in this study may serve as a basis for further assessment of the endocrine profile of women with benign and malignant breast disease.

References

1 Berle, P. and Voigt, K.D.: Evidence of plasma prolactin levels in patients with breast cancer. Am. J. Obstet. Gynec. *114:* 1101–1102 (1972).
2 Bohnet, H.G.; Dahlén, H.G.; Keller, E.; Friedrich, E.; Schindler, A.E.; Wyss, H.I., and Schneider, H.P.G.: Human pituitary gonadotrophin index. II. LH-RH test and clomiphene response before and after LH-RH stimulation. Clin. Endocrinol. *5:* 25–36 (1976).
3 Bohnet, H.G.; Mühlenstedt, D.; Hanker, J.P.F., and Schneider, H.P.G.: Prolactin oversuppression. Arch. Gynäk. *223:* 173–178 (1977).
4 Bohnet, H.G.; Gómez, F., and Friesen, H.G.: Prolactin and estrogen binding sites in the mammary gland of the lactating and non-lactating rat. Endocrinology *101:* 1111–1121 (1977).
5 Bohnet, H.G.; Pozo, E. del, and Gómez, F.: Control of mammary prolactin receptors in the female rat. Acta endocr., Copenh. *87:* suppl. 215, pp. 7–8 (1978).

6   Bohnet, H.G.; Hanker, J.P.; Horowski, R.; Wickings, E.J., and Schneider, H.P.G.: Suppression of prolactin secretion by lisuride throughout the menstrual cycle and in hyperprolactinaemic menstrual disorders. Acta endocr., Copenh. *92:* 8–19 (1979).

7   Bohnet, H.G. and McNeilly, A.S.: Prolactin: assessment of its role in the human female. Hormone metabol. Res. *11:* 533–546 (1979).

8   Bohnet, H.G.; Hanker, J.P.; Hilland, U., and Schneider, H.P.G.: Epimestrol in treatment of inadequate luteal progesterone secretion. Fertil. Steril. (in press 1980).

9   Cole, E.N.; England, P.C.; Sellwood, R.A., and Griffiths, K.: Serum prolactin concentrations throughout the menstrual cycle of normal women and patients with recent breast cancer. Eur. J. Cancer *13:* 677–684 (1977).

10 Costlow, M.E.; Buschow, R.A.; Richert, N.J., and McGuire, W.: Prolactin and estrogen binding in transplantable hormone-dependent and autonomous rat mammary carcinoma. Cancer Res. *35:* 970–974 (1975).

11 Franchimont, P.; Dowecy, G.; Légros, J.-J.; Reuter, A.; Vrindts-Gevaert, Y.; Cauwenberge, J.R. van; Remacle, P.; Gaspard, U. et Colin, C.: Dosage de la prolactine dans les conditions normales et pathologiques. Annls Endocr. *37:* 127–156 (1976).

12 Franks, S.; Ralphs, D.N.; Seagrott, V., and Jacobs, H.S.: Prolactin concentrations in patients with breast cancer. Br. med. J. *iv:* 320–321 (1974).

13 Friedrich, E.; Jaeger-Whitegiver, E.R.; Bieder, M.; Halver-Schmidt, H.; Penke, B.; Pallai, P.; Keller, E., and Schindler, A.E.: Standardisation of specific radioimmunoassays for plasma estrogen, estradiol, progesterone and androstenedione. J. Steroid Biochem. *5:* 305 (1974).

14 Grattarola, R.: The premenstrual endometrial pattern of women with breast cancer. A study of progestational activity. Cancer, N.Y. *17:* 1119–1122 (1964).

15 Henderson, B.E.: Elevated serum levels of estrogens and prolactin in daughters of patients with breast cancer. New Engl. J. Med. *293:* 790–795 (1975).

16 Hoff, J.; Hoff-Bardier, M., and Bayard, F.: Laprolactine, son rvle dans l'hormonodépendance des cancers du sein. J. Gyn. Obst. Biol. Repr. *7:* 19–30 (1978).

17 Kwa, H.G.; De Jong-Bakker, M.; Engelsman, E., and Cleton, K.: Plasma prolactin in human breast cancer. Lancet *i:* 433–434 (1974).

18 MacMahon, B.; Cole, P., and Brown, J.: Etiology of human breast cancer: a review. J. natn. Cancer Inst. *50:* 21–42 (1973).

19 Meites, J.; Lu, K.H.; Wuttke, W.; Welsch, C.W.; Nagasawa, H., and Quadri, S.K.: Recent studies on functions and control of prolactin secretion in rats. Recent Prog. Horm. Res. *28:* 471–526 (1972).

20 Mittra, I.; Hayward, J.L., and McNeilly, A.S.: Hypothalamic-pituitary-prolactin axis in breast cancer. Lancet *i:* 889–891 (1974).

21 Rolandi, E.; Barreca, T.; Masturzo, P.; Polleri, A.; Indiveri, F., and Barabino, A.: Plasma-prolactin in breast cancer. Lancet *ii:* 845–846 (1974).

22 Sherman, B.M. and Korenman, S.G.: Inadequate corpus luteum function: a pathophysiological interpretation of human breast cancer epidemiology. Cancer, N.Y. *33:* 1306–1312 (1974).

23 Sitruk-Ware, L.R.; Sterkes, N., and Mowszowicz, I.: Inadequate luteal function in women with benign breast diseases. J. clin. Endocr. Metab. *44:* 771–774 (1977).

24 Swain, M.C.; Hayward, J.L., and Bulbrook, R.D.: Plasma oestradiol and progesterone in benign breast disease. Eur. J. Cancer *9:* 553–556 (1973).

25 Vekemans, M. and Robyn, C.: The influence of exogenous estrogen on the circadian periodicity of circulating prolactin in women. J. clin. Endocr. Metab. *40:* 886–888 (1975).
26 Wilson, R.G.; Buchan, R., and Roberts, M.M.: Plasma prolactin and breast cancer. Cancer, N.Y. *33:* 1325–1337 (1974).

H.G. Bohnet, MD, Department of Obstetrics and Gynecology,
Westfalian Wilhelms-University, Westring 11, D-4400 Münster (FRG)

Discussion

Judd: When we first described the metoclopramide tests in 1976, we used a dose of 2.5 mg, which we had found produced the maximum response. By using 10 mg, you use a suprapharmacological dose. In a situation in which you are really looking for small changes, you might be better to use 0.5 mg, which provides a half-maximum response. The second problem is that, I noticed, the levels which you reached with that dose were around 5,000 μU/ml, which is at the extreme limits of the assay. What binding did you have at that levels, and with no standard errors, could you really distinguish between those things?

Bohnet: We dilute the serum before we start the assay.

Eastman: Could I ask about your controls? Were they just normal women? Where did you obtain them?

Bohnet: The controls were normal ovulating women, mostly medical staff, aged 20–45 years, which is a similar age range to the patients.

Eastman: If you were doing an animal experiment, that would not be accepted as an adequate control group. You said that the patients had their basal levels and metoclopramide tests done on the day of their admission for biopsy. That group would be under great stress, coming in for a biopsy with the obvious suspicion of breast cancer, so the control group would not be adequate. You would need to show that those levels are elevated in a comparable group coming in for, say, thyroid biopsy, rather than normal women.

Bohnet: We have repeated these metoclopramide tests in normal ovulating women and in some of these patients, and we did not see very much difference in the responses to metoclopramide.

Clinical Hyperprolactinaemia: Its Effects and Treatment

A Cross-Cultural Comparison of Prolactin Secretion in Long-Term Lactation

Barbara A. Gross, Stephen P. Haynes, Creswell J. Eastman, Virginia Balderrama-Guzman and Lilia V. del Castillo

Endocrine Unit, Woden Valley Hospital, Canberra, A.C.T.; Endocrine Unit, Department of Medicine, Westmead Centre, Westmead, N.S.W., and Department Community Health, Institute of Public Health, University of the Philippines, Manila

The role of prolactin (PRL) in the maintenance of long-term lactation and infertility in the human has not been clarified. Recent cross-sectional studies of PRL secretion in long-term lactating, urban Australian women [3, 4] and rural African women [1, 2] have revealed that a high proportion of these women has elevated basal serum PRL concentrations. Conclusive evidence for increased PRL secretion in long-term, puerperal lactation has been derived from measurement of 24-hour integrated concentrations of PRL [4]. The frequency of breast-feeding is a major influence on basal serum PRL concentration and daily PRL production rate [3, 4]. A possible causal link between PRL secretion and reproductive function during lactation has been suggested by the close correlation between resumption of menstruation and decline in PRL secretion [5].

The present study was undertaken to compare PRL secretion in two different socio-cultural populations as part of a study of factors affecting PRL secretion in lactating mothers. The study compares serum PRL levels in lactating Filipino and Australian women.

Methods and Subjects

Single blood sample were collected from 2 groups each of 50 breast-feeding mothers between one and 21 months postpartum; 50 were from Pasay City (urban Manila) and 50 were from San Rafael (a traditional rural area). Mothers and babies were observed at a study-centre over a 24-hour period as part of a parallel study to determine the 24-hour production of breast milk. Milk production was measured by test weighing and a record was kept of the frequency and duration of breast-feeding and supplementary feeding. Blood samples were taken at the end of the 24-hour period, approximately 2–3 h after the last breast-feed. Mothers and babies were healthy at the time of the study and no mother was using hormonal contraception. Sociodemographic data and reproductive histories were obtained as well as information concerning introduction of supplementary feeding and menstrual status. Blood was taken from 22 non-pregnant, non-lactating, single women (mean age 24 ± SEM 0.67) resident in urban Manila.

The method of study for the Australian sample has been described previously [3]. Basal serum prolactin was measured by a double antibody radio-immunoassay [7] using VLS-3 kit kindly donated by the National Institute of Arthritis, Metabolism and Digestive Diseases, USA. All blood samples from the Philippines were assayed in the one assay in the Australian laboratory.

Results

Sample Characteristics

The relevant demographic variables of the two population groups are summarised in table I. The mean age of the Australian women was 28.6 ± 0.7 years which was significantly higher than the mean age of 24.8 ± 0.4 years of the Filipino women ($p < 0.0005$). Parity, age of baby and frequency of breast-feeding were comparable. It is interesting to note that menstruation resumed much earlier in the Filipino women (6.3 ± 0.6 months) compared with the Australian women (11.5 ± 1.3 months). This was an unexpected and unexplained finding.

Comparisons of Serum PRL Concentrations

The mean basal serum PRL concentrations in normal Filipino and Australian women were 12.7 ± 0.9 and 10.3 ± 1.3 μg/l, respectively. The difference was not statistically significant. The mean basal serum PRL levels were elevated in the 57 amenorrhoeic Filipino women, lactating up to 20 months postpartum (fig. 1). At 1 month postpartum the mean serum PRL was 65 ± 4.4 μg/l and declined significantly to 48.6 ± 7.5 μg/l at 2 months ($p < 0.01$). Serum PRL levels continued to decline until 7 months postpartum with no further change over the next 7 months (fig. 1). The pattern was similar to, but not identical with, the pattern we have described previously in Australian women [1]. Serum PRL levels in menstruating lactating Filipino women from 3 to 22 months postpartum are shown in figure 2.

The mean basal serum PRL concentration in amenorrhoeic lactating women was significantly higher than the concentration in menstruating, lactating women in both the Filipino ($p < 0.001$) and Australian groups

A Cross-Cultural Comparison of Prolactin Secretion in Long-Term Lactation   181

Table I. A comparison of demographic factors and breast-feeding frequency in a Filipino and an Australian sample of breast-feeding women (means ± SEM)

Characteristic	Philippines (n=99)	Australia (n=29)
Age, years	24.8 ± 0.4	28.6 ± 0.7*
Parity	2.4 ± 0.2	2.4 ± 0.1
Age of baby, months	7.8 ± 0.5	7.3 ± 0.6
Number of breast-feeds in 24 h	7.8 ± 0.2	7.6 ± 0.5
Duration of amenorrhoea, months	6.3 ± 0.6 (n=49)	11.5 ± 1.3

* $p < 0.0005$.

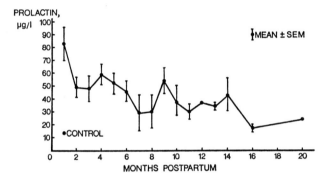

Fig. 1. Mean basal serum PRL levels in lactating amenorrhoeic Filipino women (n=57) from 1 to 20 months postpartum. (A cross-sectional study.)

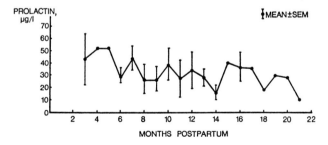

Fig. 2. Mean basal serum PRL levels in lactating menstruating Filipino women (n=41) from 3 to 22 months postpartum. (A cross-sectional study.)

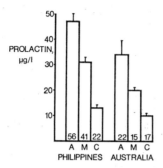

Fig. 3. A comparison of mean basal serum PRL levels in lactating amenorrhoeic (A), lactating menstruating (M) and non-pregnant, non-lactating (C) women in a sample of Filipino and Australian women. The number of women studied in each group is shown in the columns.

Table II. A comparison of mean basal serum PRL levels in non-pregnant, non-lactating women, amenorrhoeic and menstruating lactating women and amenorrhoeic, full and partial breast-feeding women in a Philippine and an Australian sample (means ± SEM)

Subjects	Philippines		Australia		Significance level
	PRL µg/l	n	PRL µg/l	n	
Normal controls	12.7 ± 0.9	22	10.3 ± 1.3	17	n.s.
Amenorrhoeic	46.9 ± 3.0	56	33.8 ± 5.4	22	$p < 0.05$
Menstruating	30.6 ± 2.4	41	20.2 ± 0.9	15	$p < 0.02$
Amenorrhoeic					
Full breast-feeding	57.9 ± 5.1	16	50.1 ± 11.1	9	n.s.
Partial breast-feeding	42.2 ± 3.1	38	22.5 ± 2.2	13	$p < 0.001$
Partial					
Milk formula or juice only	53.5 ± 5.9	10	—		n.s.*
Solids and liquids	38.1 ± 3.4	28	22.5 ± 2.2	13	$p < 0.01$

* As compared to fully breast-feeding Filipino or Australian women.

($p < 0.05$; fig. 3). The lactating menstruating women had significantly higher serum PRL levels than the normal controls ($p < 0.001$). These data are shown in table II.

When the Filipino and Australian women were compared, statistically significant differences were noted between each comparable group, except for the normal controls. In the amenorrhoeic women, mean serum PRL in

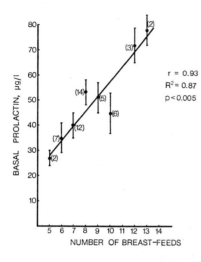

Fig. 4. Correlation between frequency of breast-feeding and serum PRL concentration.

the Filipino group was $46.9 \pm 3.0\,\mu g/l$ compared with a mean level of $33.8 \pm 5.4\,\mu g/l$ for Australian women ($p < 0.05$). Similarly, in the menstruating women, the mean level of $30.6 \pm 2.4\,\mu g/l$ in the Filipino group was significantly higher than the mean level of $20.2 \pm 0.9\,\mu g/l$ in the Australian group ($p < 0.02$) (table II).

Full and Partial Breast-Feeding Comparisons in Amenorrhoeic Women

We reanalysed the data for amenorrhoeic women to determine the effect of full and partial breast-feeding on serum PRL concentration (table II). The mean serum PRL level in fully breast-feeding women in the Philippines ($57.9 \pm 5.1\,\mu g/l$) was not significantly different from the mean level of $50.1 \pm 11.1\,\mu g/l$ in Australian women. Analysis of the data from partially breast-feeding women revealed significant differences between the two ethnic groups.

While the Filipino women showed a small but significant decline from $57.9 \pm 5.1\,\mu g/l$ in fully breast-feeding to $42.2 \pm 3.1\,\mu g/l$ in partially breast-feeding ($p < 0.01$), the decline in Australian women from 50.1 ± 11.1 to $22.5 \pm 2.2\,\mu g/l$ was more marked ($p < 0.01$). Thus, it would appear that the differences in serum PRL in the overall amenorrhoeic groups can be attributed to the greater decline in serum PRL level occurring with partial breast-feeding in Australian women. This conclusion is consistent with the results observed after analysis of serum PRL levels in partial breast-feeders classified according to the nature of the feeding supplement (table II).

Breast-Feeding Frequency and Serum PRL Concentration

There was a highly significant correlation between the frequency of breast-feeding and serum PRL in both the amenorrhoeic lactating and menstruating lactating Filipino women ($r = 0.83$, $p < 0.005$) (fig. 4). These data were not available for Australian women.

Discussion

Although the mean age of the lactating Australian women was greater than the Filipino women the two groups were well matched for parity and age of baby. It is of interest that the frequency of breast-feeds per 24 h was almost identical in both population groups. There are few data available for comparison of breast-feeding frequencies in these population groups. The ability of women to maintain lactation long-term may, in large part, be attributable to the frequency of breast-feeding. An important difference between the two groups was the duration of amenorrhoea. The mean duration of amenorrhoea of 11.5 months in the Australian group was significantly longer than the 6.3 months in the Filipino women. This finding was the reverse of what might be predicted after reference to serum PRL levels and the quality of nutrition in the two groups. The explanation for this finding is obscure.

There was no significant difference in serum PRL levels between the normal Filipino and Australian control groups. Similarly, the fully breast-feeding, amenorrhoeic women exhibited comparable elevations in serum PRL concentrations. The pattern of serum PRL levels in the lactating Filipino women was similar to what we have described previously in lactating Australian women [3] and the report of *Delvoye et al.* [1] in African women. Thus, it is highly likely that increasing PRL secretion is a universal finding in long-term lactation in humans.

Comparison of serum PRL levels between lactating amenorrhoeic women and lactating menstruating women revealed significant differences. In both population groups, serum PRL was significantly higher in the amenorrhoeic women. These results confirm our previous report [3] and that of *Delvoye et al.* [1]. It is interesting to note that the decline in serum PRL in the partial breast-feeding groups was significantly greater in the Australian women compared with the Filipino women. The reason for this is not certain, however, it is likely that the result is a function of time. In general, the Filipino mothers introduced supplementary feeds earlier than their Australian counterparts, 77% of the Filipino women had commenced fluid supplements by 3 months postpartum. This early introduction of fluids did not result in a significant decrease in PRL levels in mothers who were studied in the first 3 months postpartum compared to fully breast-feeding women, although there were significantly lower levels in individual women who had also resumed menses by 3 and 4 months postpartum. PRL levels were significantly lower in mothers, in both the Filipino and Australian samples, who had introduced solid foods (at a mean of 4.5 months in the Philippines and 5 months in Australia), but mean PRL levels in the Filipino women were still significantly higher than the Australian women who were partially breast-feeding.

These studies confirm that the basal serum PRL level provides only a crude index of menstrual status in an individual woman. Although a serum PRL level in or near the upper limit of the normal range would indicate that a mother had resumed or was about to resume menstruation, an elevated serum PRL level was found in many mothers who had resumed menstruation. Whether or not ovulation and normal luteal function occurred in the menstruating women remains to be determined. Inadequate luteal function is a common occurrence in lactating Australian women returning to menstruation [6].

Acknowledgements

We are grateful to the Philippine research assistants and in particular *Elizabeth Espiritu* for the collection of data in the Philippines, and *Freda Doy* for expert technical assistance. This work was supported by a Ford-Rockefeller research grant to *Barbara A. Gross* and *Creswell J. Eastman.*

References

1 Delvoye, P.; Desnoeck-Delogne, J., and Robyn, C.: Serum prolactin in long lasting amenorrhoea. Lancet *ii:* 288–289 (1976).
2 Delvoye, P.; Demaegd, J.; Delogne-Desnoeck, J., and Robyn, C.: The influence of the frequency of nursing and of previous lactation experience on serum prolactin in lactating mothers. J. biosoc. Sci. *9:* 447–451 (1977).
3 Gross, B.A. and Eastman, C.J.: Prolactin secretion during prolonged lactational amenorrhoea. Aust. N.Z. J. Obstet. Gynaec. *19:* 95–99 (1979).
4 Gross, B.A.; Eastman, C.J.; Bowen, K.M., and McElduff, A.P.: Integrated concentrations of prolactin in breastfeeding mothers. Aust. N.Z. J. Obstet. Gynaec. *19:* 150–153 (1979).
5 Gross, B.A.; Eastman, C.J.; Haynes, S.P., and McElduff, A.P.: A prospective study of prolactin, gonadotrophin and ovarian steroid secretion during prolonged lactation amenorrhoea. 6th Int. Congr. of Endocrinology, Melbourne 1980, abstr. No. 651.

6 Gross, B.A.; Haynes, S.P., and Eastman, C.J.: The inadequate luteal phase in lactational hyperprolactinaemia (in preparation, 1980).
7 Sinha, Y.N.; Selby, F.W.; Lewis, U.J., and Laan, W.P. vander: A homologous radioimmunoassay for human prolactin. J. clin. Endocr. Metab. *36:* 509–516 (1973).

Dr. C.J. Eastman, Endocrine Unit, Division of Medicine, Westmead Medical Centre, Westmead, N.S.W. 2145 (Australia)

Discussion

Robyn: In lactating mothers from Kivu serum levels of prolactin are remaining high for 18–20 months postpartum. In the population of a European country, where breast-feeding is much less frequent, because, for example, the child does not sleep with the mother and the child is not with the mother all day, and breast-feeding is carried out with six or seven suckling episodes per day, whereas in Africa it is around 12 suckling episodes per day, there is a more dramatic fall: after 8 or 9 months the levels are close to the normal range.

Of course, stimulation of the breast by suckling is important, but one has also to recall that in some of these countries, at least in Africa, they use infusions of plants when breast-feeding is not very successful. We were surprised to see that some of these plant infusions had definite effects on prolactin levels. One should, perhaps, be careful about the origins of hyperprolactinaemia in these lactating women. One should investigate a little more closely what kinds of natural things can stimulate prolactin.

Eastman: One would be most interested to know what was in the plants.

Judd: When were your basal prolactin levels taken, in relation to the last breast-feed and in relation to the time of day?

Gross: We took almost all of our basal samples, in both the Philippine group and the Australian group, at least two and preferably 3 h after a feed. We found that the time after a feed is critical, because if you take it too close you will have an elevated prolactin level. The majority of samples were also taken between 0900 and 1100 in the morning. In the Philippines, in particular, we were able to take this following the 24-hour observation of milk production and feeding habits.

Robyn: It is quite obvious that after suckling there is a rise that is significant. But it is not a very marked increase and certainly not as seen in the very early stages of lactation. There has been some indication that the mechanism by which prolactin levels are maintained at high during lactation is not only due to the suckling effect; after this direct effect there is another mechanism which is responsible for maintaining the high level. One has to keep clear that the levels are not just going up and down during the day; on average they are staying high, even a long time after a suckling episode.

Gross: I agree with you on that. There is fairly good evidence to show that suckling has an effect, but that is very variable. Even when we take a control sample 2 h after a feed, it might not be back down to the normal basal level. It can remain elevated, and it can continue to go up after each feed.

Eastman: The response to suckling is variable, and not necessarily reproducible. You can produce a very marked increase in 24-hour production, often with apparently normal basal levels. This is why, in the past, people have missed the point of prolactin in relation to suckling; the studies have not been carried out in the proper manner.

Role of Transient Hyperprolactinaemia in the Short Luteal Phase

J.R.T. Coutts, R. Fleming, A. Craig, D. Barlow, P. England and M.C. Macnaughton

Department of Obstetrics and Gynaecology, Glasgow University; Clinpath Services Limited, High Wycombe, Bucks., and Department of Pathological Biochemistry, Glasgow Royal Infirmary

No functional role for prolactin (PRL) has yet been clearly established in the normal human menstrual cycle. Although cyclical changes in peripheral plasma concentrations of PRL, usually correlated with circulating oestradiol (E_2) levels, have been observed [5], no particular event appears crucial to the functioning of the menstrual cycle. Conversely, a functional role for PRL in the normal cycle was suggested by the demonstration that human luteinised granulosa cells *in culture* secreted significantly less progesterone (P) in the presence of an antiserum to PRL [8] and also that human follicular fluid apparently selectively concentrated PRL [9].

Hypersecretion of PRL has been clearly linked with disruptions of the menstrual cycle [1]. Normalisation of PRL levels in patients with amenorrhoea-galactorrhoea, using bromocryptine, usually produces normal cycles and often restores fertility [15]. Hyperprolactinaemia may cause anovulation through its effects either at the hypothalamic/pituitary level by affecting gonadotrophin biosynthesis or release [6] or at the ovarian level by inhibiting follicular growth and development [13] or both.

By daily assessment throughout menstrual cycles of infertile women we have identified a condition which has been defined as transient hyperprolactinaemia (T↑PRL) [4]. This T↑PRL was observed as a discrete event of variable duration at different stages of the cycle. Since there was no pharmacological cause of the elevated PRL levels, these patients provided a good group for studying any effects of PRL on the menstrual cycle. To further investigate any associations between PRL levels and the menstrual cycle we have studied the effects of metoclopramide-induced T↑PRL in volunteers with normal menstrual history.

Methods

Blood Samples

Serial daily blood samples were obtained throughout a single menstrual cycle from 32 infertile women with normal menstrual rhythm and from 7 normally cycling women. A further 6 normally cycling volunteers gave serial daily samples throughout two consecutive cycles; the first was a control cycle and in the second metoclopramide (Maxolon, Beecham Research Laboratories, 10 mg, 3 × daily) was administered for 6 days over the estimated mid-cycle period – sufficient to maintain hyperprolactinaemia over 24 h for the full 6-day period [7].

Hormone Assays

Using radio-immunoassays the levels of PRL, LH, FSH, E_2 and P were determined in all samples. In the case of the metoclopramide study, to avoid errors due to inter-assay variation, all samples were analysed in the same assay for protein hormones and the samples from both luteal phases from each subject were assayed simultaneously for steroid hormones.

We divided the cycle into 4 compartments [2]: follicular phase (days –12 to –3 inclusive); mid-cycle (–2 to +2 inclusive); early luteal phase (days +1 to +8 inclusive for gonadotrophins and PRL; days +3 to +8 inclusive for steroid hormones) and late luteal phase (days +9 to the end of the cycle inclusive). 'Total' steroid hormone levels were determined by summing all of the respective E_2 or P values within that phase whilst PRL, LH and FSH values were expressed as mean levels within each phase.

Results

Daily plasma sampling throughout the cycle led to the diagnosis of T↑PRL in 21 of the 32 patients with unexplained infertility. Linear correlation analysis of these data showed no direct correlations between mean PRL levels and the levels of any of the other hormones within any phase of the cycle. However, significant indirect negative correlations were observed between mean mid-cycle PRL levels and late luteal phase 'total' levels of both E_2 ($r = -0.74$) and P ($r = -0.44$).

Figure 1 demonstrates the negative correlations between mean mid-cycle PRL levels and total late luteal E_2 (fig. 1a) and P (fig. 1b) levels. The levels of the steroid hormones were significantly lowered (P: $p < 0.01$; E_2: $p < 0.001$) when mean mid-cycle PRL levels were above the normal upper limit (580 mIU/l). These results suggested an association between mid-cycle hyperprolactinaemia and luteal phase length which was confirmed by examining the relationship between mean mid-cycle PRL levels and luteal phase length. When mean mid-cycle PRL levels were above the normal values, the lengths of the luteal phases were significantly shorter ($p < 0.001$).

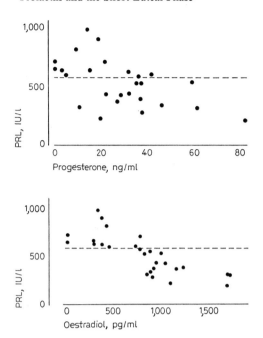

Fig. 1. Effect of mean mid-cycle prolactin levels on 'total' late luteal phase steroid hormone levels.

The mean hormone profiles from 9 cycles with short luteal phases (defined as < 11 days from LH peak to the day when the P level fell to < 1.5 ng/ml) are shown in figure 2a, b. These profiles show elevated PRL levels over the mid-cycle period with somewhat normal gonadotrophins and generally normal steroid hormone levels which were truncated earlier than usual.

Using metoclopramide, 4 volunteers showed consequently elevated PRL levels throughout the mid-cycle period; in another subject the PRL levels were elevated for part of the mid-cycle period (days −6 to −2 inclusive) whilst in the final volunteer the levels were elevated during the luteal phase (days +2 to +6 inclusive). PRL concentrations ranged from 750 to 5,360m IU/l (mean: 2,495 ± SD 1,376 m IU/l). Analysis of the LH and FSH values during the peak (mid-cycle) phase, using a t-test for paired data, revealed a significant reduction ($p < 0.01$) of the LH values on metoclopramide, but the positive feedback of gonadotrophins was certainly not abolished. 'Total' late luteal phase steroid production, as estimated by summing the concentrations of respectively E_2 and P observed from days +9 to the end of the

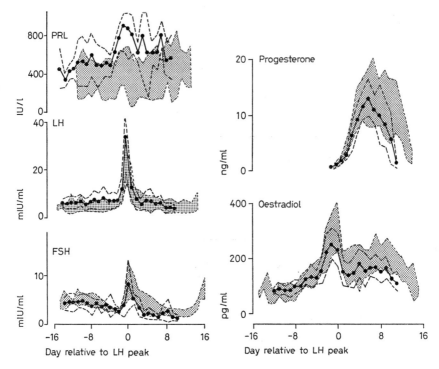

Fig. 2. Mean (●—●) ± SD (– –) profiles of LH, FSH, prolactin, oestradiol and progesterone in patients with short luteal phases (n=9) compared with the 95% confidence limits of the normal cycles (stippled background).

cycle, were not significantly different in either case between the metoclopramide-treated and control cycles. Table I lists the length of the luteal phases in control and metoclopramide-treated cycles in the 5 patients in whom midcycle T↑PRL was induced. No significant difference was observed between luteal phase lengths in the two groups of cycles.

Discussion

Short luteal phases in women have been recognised as discrete hormonal pathologies for many years [10, 12, 14]. It has been suggested that the condition might be a consequence of altered LH/FSH ratios [14] and more recently hyperprolactinaemia has been implicated in the short luteal phase by observations following withdrawal of bromocryptine treatment from

Table I. Effect of mid-cycle metoclopramide-T↑PRL on luteal phase length

	Luteal phase length days
Control cycle (n = 5)	12.6 ± SD 1.5
Metoclopramide treatment cycle (n = 5)	12.4 ± SD 1.9

hyperprolactinaemic women [11]. The data presented in this paper showed a significant association between T↑PRL occurring at mid-cycle and both late luteal phase steroid production and luteal phase length. No direct effects of elevated PRL levels within any phase of the cycle affecting ovarian steroidogenesis within that phase of the cycle were observed.

Attempts to induce short luteal phases by metoclopramide-stimulated mid-cycle T↑PRL were unsuccessful. Previously *Delvoye et al.* [3] showed that sulpiride-induced hyperprolactinaemia (over a prolonged period) in normally menstruating women eliminated or reduced both the mid-cycle positive feedback peaks of LH and FSH and also luteal phase P levels. Luteal phase length was also shortened by the sulpiride treatment which had been initiated in the late follicular phase. In contrast the metoclopramide induced T↑PRL observed in this study did not severely inhibit the positive feedback; neither follicular maturation (represented by pre-ovulatory E_2 levels) nor luteal phase steroid hormone profiles were affected. The present results indicate that the failure of the positive feedback observed in hyperprolactinaemic anovulatory patients [6] might not be directly due to the immediate circulating PRL levels and also that PRL *per se* probably does not have major direct effects on ovarian steroid hormone production. The association observed between mid-cycle T↑PRL and short luteal phases may therefore be an indirect relationship reflecting on the aetiology of the hyperprolactinaemia rather than as a consequence thereof.

References

1 Bohnet, H.G.; Dahlen, H.G.; Wuttke, W., and Schneider, H.P.G.: The hyperprolactinaemic anovulatory syndrome. J. clin. Endocr. Metab. *42:* 132–143 (1976).
2 Coutts, J.R.T.; Fleming, R.; Carswell, W.; Black, W.P.; England, P.; Craig, A., and Macnaughton, M.C.: The defective luteal phase; in Jacobs, Advances in gynaecological endocrinology, pp. 65–101 (RCOG, London 1978).

3   Delvoye, P.; Taubert, H.D.; Jurgensen, O.; L'Hermite, M.; Delogne, J., and Robyn, C.: Influence of circulating prolactin increased by a psychotropic drug on gonadotrophin and progesterone secretion. Acta endocr., Copenh. suppl. 184, p. 110 (1974).
4   Fleming, R.; Craig, A.; England, P.; Macnaughton, M.C., and Coutts, J.R.T.: Influence of hyperprolactinaemia on ovarian function in infertile women with normal menstrual rhythm. J. Endocr. 77: 14P–15P (1978).
5   Franchimont, P.; Dourcy, C.; Legros, J.J.; Reuter, A.; Vrindts-Gevaert, Y.; Van Cauwenberge, J.R., and Gaspard, U.: Prolactin levels during the menstrual cycle. Clin. Endocrinol. 5: 643–650 (1976).
6   Glass, M.R.; Shaw, R.W.; Williams, J.W.; Butt, W.R.; Logan-Edwards, R., and London, D.R.: The control of gonadotrophin release in women with hyperprolactinaemic amenorrhoea: effect of oestrogens and progesterone on the LH and FSH response to LHRH. Clin. Endocrinol. 5: 521–530 (1977).
7   Healy, D.L. and Burger, H.G.: Increased prolactin and thyrotrophin secretion following oral metoclopramide; dose response relationships. Clin. Endocr. 7: 195–201 (1977).
8   McNatty, K.P.; Hunter, W.N.; McNeilly, A.S., and Sawers, R.S.: Changes in the concentrations of pituitary and steroid hormones in the follicular fluid from human Graafian follicles throughout the menstrual cycle. J. Endocr. 64: 555–571 (1975).
9   McNatty, K.P.; Sawers, R.S., and McNeilly, A.S.: A possible role for prolactin in control of steroid secretion by human Graafian follicles. Nature, Lond. 250: 653–655 (1974).
10 Ross, G.T.; Cargille, C.M.; Lipsett, M.B.; Rayford, P.L.; Marshall, J.R.; Strott, C.A., and Rodbard, D.: Pituitary and gonadal hormones in women during spontaneous and induced ovulatory cycles. Recent Prog. Horm. Res. 26: 12–48 (1970).
11 Seppälä, M.; Hirvonen, E., and Ranta, T.: Hyperprolactinaemia and luteal insufficiency. Lancet i: 229–230 (1976).
12 Sherman, B.M. and Korenman, S.G.: Measurement of plasma LH, FSH, estradiol and progesterone in disorders of the human menstrual cycle: the short luteal phase. J. clin. Endocr. Metab. 38: 89–93 (1974).
13 Strauch, G.; Bonnefous, S.; Paulian, B.; Zaks, P.; Pages, J.P., and Bricaire, H.: Studies on the epidemiology and mechanisms of hyperprolactinaemic anovulation; in Crosignani and Robyn, Prolactin and human reproduction, pp. 161–175 (Academic Press, New York 1977).
14 Strott, C.A.; Cargille, C.M.; Ross, G.T., and Lipsett, M.B.: The short luteal phase. J. clin. Endocr. 30: 246–251 (1970).
15 Thorner, M.O.; McNeilly, A.S.; Hagan, C., and Besser, G.M.: Long-term treatment of galactorrhoea and hypogonadism with bromocriptine. Br. med. J. ii: 419–422 (1974).

J.R.T. Coutts, BSc, PhD, Department of Obstetrics and Gynaecology, Glasgow University, Glasgow (Scotland)

Discussion

Judd: When patients were given bromocriptine and ovarian cyclicity restored and then bromocriptine was stopped during part of the normal menstrual cycle, it had to be stopped within 5 days of the start of the cycle to cause any change in basal body temperature, implying that the effect of prolactin in the early follicular phase resulted in the luteal effects. I wonder whether *Coutts* might have just missed the boat in giving just the metoclopramide at mid-cycle. Perhaps, if he had given it throughout the cycle, he might have noticed a change.

Coutts: That is probably a fair comment, and I think that *Robyn* is going to show that they get something similar from their data. But from our data, when we attempted to do correlations between the early follicular phase prolactin levels and luteal phase progesterone levels, we could find no correlation.

Haines: The inadequate luteal phase is well-associated with the pathological and pharmacological hyperprolactinaemia. In lactational amenorrhoea, we have observed that the return of fertility can be preceded by inadequate luteal activity in cycles subsequent to the return of menstruation. As part of a prospective longitudinal study of lactational amenorrhoea, by *Gross*, it was observed that 16 of 33 breast-feeding mothers had at least one inadequate luteal phase after the return of menses and prior to the return of fertility. There is a significant fall in prolactin levels over these first three cycles. As a point of interest, the serum prolactin levels in the group of patients who returned to fertility with normal luteal function, compared to the group who returned with inadequate luteal function, is highly significant, a level of 13.8 compared to 24.8.

Coutts: High prolactin levels do exist in women with short luteal phases; we have shown that on a statistically significant basis. The 'take-home message' is that the prolactin levels at mid-cycle are probably not causing the short luteal phase, some changes in dopamine turnover are occurring which result in high prolactin levels.

Robyn: We tried to induce hyperprolactinaemia over short periods during the follicular phase (from day 1 to day 10 of the cycle). All the events of the cycle were normal. When inducing hyperprolactinaemia from day 10 of the cycle to day 17, there was no rapid follicular growth but the rise in follicular oestrogens occurred after the interruption of treatment. This was followed by a normal LH surge and a normal luteal phase. From these data, we can say that, to get a change in the cycle by hyperprolactinaemia, one has to induce it for rather a long period, and this has to include the period of rapid follicular growth. Without that, we do not see changes in developing granulosa cells.

Medical Therapy of Prolactin-Secreting Pituitary Tumours

Sven Johan Nillius

Department of Obstetrics and Gynaecology, University Hospital, Uppsala

Prolactin-secreting pituitary adenomas (prolactinomas) are very common causes of pathological hyperprolactinaemia. The clinical features o- increased prolactin secretion are inappropriate lactation and gonadal dysf function. Galactorrhoea is not an obligatory symptom even in women; it is uncommon in men. The effects of hyperprolactinaemia on gonadal function are clinically much more important.

Amenorrhoea is the most common symptom of hyperprolactinaemia in *women*. The prevalence of hyperprolactinaemia in amenorrhoea ranged between 13 and 39% in 8 large series of totally 1,305 patients [2]. *Rjosk et al.* [80] recently reported that 18.8% of 750 women with at least 6 months of amenorrhoea had raised prolactin levels and one third of them had radiological changes of the sella turcica. This agrees well with other reports in the literature [2, 56, 83]. Radiographic sellar abnormalities consistent with a pituitary tumour are very frequent in amenorrhoeic women with hyperprolactinaemia.

In *men*, hyperprolactinaemia is much less common. The typical findings in men with hyperprolactinaemia are clinical and biochemical evidence of testosterone deficiency with decreased or absent libido and impotence. In two series of totally 156 men with sexual impotence, hyperprolactinaemia was diagnosed in 0 and 7.8%, respectively [10, 79]. In men presenting with infertility, raised prolactin levels are not frequently observed. The prevalence of hyperprolactinaemia was calculated to be 3.6% in six large series of totally 941 men investigated for infertility [23, 47, 52, 79, 81, 82]. The majority of men with hyperprolactinaemia described in the literature have had large prolactinomas [1, 11, 35, 70, 95].

Medical therapy with potent dopamine receptor agonists effectively lowers raised serum prolactin levels in physiological, pharmacological and pathological hyperprolactinaemia. The ergot alkaloid bromocriptine was the first drug of this important series of therapeutic substances which became available for clinical studies [32, 51]. Bromocriptine inhibits prolactin secretion by a direct action on the prolactin-secreting cells of the pituitary [33]. It also reduces dopamine turnover in the tubero-infundibular dopamine neurons involved in the inhibitory control of prolactin secretion, suggesting a hypothalamic site, too. Bromocriptine acts by stimulating dopamine receptors [18, 50]. During recent years, other dopamine agonists like lisuride [21, 22, 48, 61], and serotonin antagonists with dopaminergic properties, like metergoline [13, 19, 31], have been introduced. They also seem to be effective in normalizing hyperprolactinaemia but the clinical experience with these drugs is still limited.

After successful lowering of raised prolactin concentrations by bromocriptine, gonadal function is restored to normal. In men with prolactin-secreting pituitary tumours, bromocriptine therapy has been found to cause a rise in testosterone secretion and to improve potency [11, 35, 88]. In women with hyperprolactinaemia, bromocriptine therapy has been extremely effective in normalizing the menstrual dysfunction. Ovulation and fertility is rapidly restored to normal in hyperprolactinaemic women with amenorrhoea [4, 34, 36, 75, 88, 89]. However, pregnancy may cause pituitary tumour complications in these women, many of whom harbour small pituitary microadenomas. The optimum management of infertile patients with hyperprolactinaemia and suspected pituitary tumour has become a burning issue during recent years. This urgent problem is reviewed and discussed below. It seems that primary medical treatment with bromocriptine is a safe and effective therapeutic alternative to surgery and radiotherapy in many infertile women with prolactinomas.

During the last decade there has been a surge of interest in transsphenoidal pituitary microsurgery [42, 45]. An increasing number of patients with prolactinomas have been submitted to surgical treatment with selective adenomectomy via the transsphenoidal approach. In patients with prolactin-secreting pituitary microadenomas, experienced neurosurgeons have produced excellent clinical and biological results with very few surgical complications [46, 55, 76]. However, the percentage of cure falls to 50% or less in patients with prolactin-secreting macroadenomas and few patients with large invasive prolactinomas are cured by surgery [30, 46, 76]. The poor surgical results in large prolactinomas have increased the interest in medical therapy. Bromocriptine has documented its value for secondary treatment of surgical failures. However, dopamine agonists may also be useful for primary treatment of prolactinomas. During the last few years, several reports on pituitary tumour regression during bromocriptine therapy have appeared in the literature. These studies will be reviewed in the second part of this paper. Medical therapy with dopamine agonists like bromocriptine seems to be of great potential value in the management of patients with prolactin-secreting pituitary tumours.

Prolactinomas and Pregnancy

The management of infertile women with hyperprolactinaemia and radiological evidence of pituitary tumour is one of the most controversial topics in clinical endocrinology today. Women with prolactin-secreting pituitary tumours usually have amenorrhoea and are infertile. Bromocriptine treatment has proved to be very effective in restoring ovulation and fertility also in women with prolactinomas [4, 87]. Ovulation can be induced by prolonged treatment with high doses of bromocriptine even in patients with large pituitary tumours and severe gonadotrophin deficiency. However, it soon became apparent that pregnancy induced by bromocriptine in patients with prolactinomas was associated with a risk of rapid tumour expansion [3].

Visual Complications during Pregnancy

Pituitary tumours may rapidly enlarge during pregnancy and cause severe headache, diabetes insipidus or serious visual complications. Tumour expansion into the suprasellar region can cause optic chiasma compression with visual field defects and decreased visual acuity. The first description of a visual complication during pregnancy appeared in 1908 [78]: *Von Reuss* reported on a 35-year-old patient, who during her 15th pregnancy had a rapid decrease in visual acuity and developed bitemporal hemianopsia, which improved after delivery. The visual disturbance deteriorated during her next pregnancy but improved after a therapeutic abortion. *Von Reuss* discussed a pituitary aetiology for the visual complication. He ended his report by commenting that if this patient ever returned to him, he would not forget to perform an X-ray of the sella turcica. Many descriptions of visual field defects developing during pregnancy in women with pituitary tumours have appeared in the literature since then. Figure 1 illustrates the severe impairment of vision that can occur during pregnancy in a patient with a pituitary tumour, but also the rapid and complete restitution after delivery.

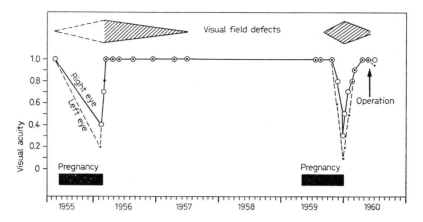

Fig. 1. Bitemporal hemianopsia and severe deterioration of vision during two successive pregnancies in a woman with a chromophobe adenoma. From *Enoksson et al.* [26], with permission.

In the last 4 years there have been at least 23 reports on tumour complications during pregnancy in patients with prolactinomas [for references, see 72]. Thus, there is an undoubted risk that visual complications may occur during pregnancy in women with pituitary tumours. The visual symptoms are apparently due to rapid growth of the adenoma or the normal pituitary gland or both during pregnancy.

Effects of Pregnancy on the Pituitary

The classical description of changes in the pituitary during pregnancy was given by *Erdheim and Stumme* [27] in 1909. They found that there was a marked increase in the size and the weight of the normal pituitary which sometimes nearly doubled in weight during pregnancy. The pituitary enlargement was due to an increase in the number of large chromophobe cells, 'schwangerschaftzellen'. These 'pregnancy cells' have later been identified as the lactotropes which produce prolactin [39, 73]. Thus, the increase in size of the pituitary during pregnancy is the result of prolactin cell hyperplasia and hypertrophy.

Many hormonal changes occur during pregnancy. There is for example a dramatic increase of the oestrogen secretion from the ovary and placenta and oestrogens are potent stimulators of prolactin secretion. During pregnancy, there is a close parallelism between the rising oestrogen, prolactin secretion and pituitary weight. It is likely that the large amounts of circulating oestrogens during pregnancy stimulate the growth of lactotropes in the normal pituitary gland and in prolactin cell adenomas, and this may give rise to complications from pressure on neighbouring structures.

Effects of Pregnancy on Pituitary Tumours

In a recent review of previously published reports and responses to a questionnaire, *Magyar and Marshall* [64] examined effects of pregnancy on previously untreated pituitary tumours. Data were obtained from 73 women who had experienced 91 pregnancies. All but 8 of these pregnancies had occurred after treatment with bromocriptine (46%), human gonadotrophins (25%) or other ovulation-inducing agents. Headache was experienced in 23% of these pregnancies, visual complications in 25%, but 61% remained asymptomatic. In those patients who developed symptoms, the median time to headache was 10 weeks and to visual disturbance 14 weeks. Surgical or radiation treatment of tumours either during or soon after pregnancy was given to 30% of the women. No patient had serious permanent sequelae. In 78 pregnancies occurring in 73 women with previously treated pituitary tumours, headache occurred in 4% and visual disturbances in 5%. 1 patient required therapy. After considering the information, these authors concluded that small pituitary tumours do not constitute a contraindication to either induction of ovulation or pregnancy.

More recently, *Gemzell and Wang* [37] reviewed the literature and surveyed the opinions of physicians around the world on effects of pregnancy on prolactin-producing pituitary adenomas. 91 pregnancies had occurred in 85 women with previously untreated prolactin-producing *microadenomas*. Ovulation had occurred spontaneously in only 2 women. Bromocriptine had been used to induce ovulation and pregnancy in 70%. Of these pregnancies, 94.5% were uneventful. Only 1 of the 85 women had a visual disturbance and this was treated conservatively during the triplet pregnancy. In contrast, 25% of the 46 women with previously untreated prolactin-producing *macroadenomas* had visual disturbances during pregnancy. The onset of the symptoms had already occurred in the first trimester of pregnancy in no less than 45% of the patients. 6 women received surgical and radiation therapy of the pituitary tumour during pregnancy. For women with previously treated pituitary adenomas (n = 67), the risk of developing complications during pregnancy was found to be 7.1%. After considering their data, *Gemzell and Wang* [37] proposed that ovulation may be induced with bromocriptine or human gonadotrophins in infertile women with prolactin-producing pituitary microadenomas. Women with pituitary macroadenomas should be operated upon or irradiated before attempting pregnancy. Ovulation may be induced with bromocriptine or human gonadotrophins postoperatively or after irradiation.

Management of Infertile Women with Prolactinomas

Until recently most authorities recommended that women with prolactin-producing pituitary tumours should be treated by either surgery or irradiation before attempting a pregnancy. It is often necessary to combine these treatments with ovulation-inducing therapy with bromocriptine or, in case of postoperative hypopituitarism, human gonadotrophins. Primary medical therapy with bromocriptine was suggested by *Bergh et al.* [3] to be a suitable treatment for many infertile women with prolactinomas. Increasing support for this therapeutic alternative has been obtained during the last few years. The possible treatments before pregnancy are reviewed below.

Surgery

A summary of published surgical results of treatment of hyperprolactinaemia was recently put together by *Thorner et al.* [90]. A wide range of results after transsphenoidal surgery was found in 8 series of a total of 283 patients (24–78% cured). The highest cure rate was described by *Franks et al.* [34] in a small group of 9 patients. *Nabarro and Franks* [69] recently updated their results in 27 patients treated by transsphenoidal surgery and then found a cure rate of 55%. 2 of their 3 patients who developed postoperative hypopituitarism were infertile and required treatment with human gonadotrophins. *Fahlbusch* [30] recently reviewed results after surgical treatment of 366 prolactinomas in 366 patients operated by experienced neurosurgeons in Milan, Munich and Paris. Microadenomas were removed selectively by transsphenoidal microsurgery in 111 patients. Hyperprolactinaemia persisted in 40% of these patients 2–3 months after the operation. After 1 year, another 7 patients had elevated prolactin levels in serum. In the whole series, the overall *failure* rate was 68%.

Pregnancy rates are not given in any of the surgical series, with one exception. In the large series of 120 operated prolactinomas, *Derome et al.* [24] show an overall pregnancy rate of 42%. Few data on the clinical course and outcome of pregnancies in patients operated on for prolactinomas are available. There is also need of more detailed information on postoperative endocrine function and on recurrence rates on long-term follow-up.

The importance of the surgeon's personal operative experience for the surgical outcome was stressed by *Wiebe et al.* [97]. These authors illustrate this by showing results after transsphenoidal surgery in 13 patients with prolactin-secreting microadenomas. Only 5 of these 13 cases became normoprolactinaemic and only 2 of the 7 infertile women conceived. One of them required addition of clomiphene. The choice of surgery as primary treatment for infertile women with prolactinomas must clearly depend on the availability of a local surgeon well experienced in transsphenoidal pituitary microsurgery.

Radiotherapy

External and interstitial pituitary irradiation have been proposed as prophylactic treatments to prevent swelling of the pituitary during pregnancy in infertile women with radiologically obvious pituitary tumours [12, 87]. Radiotherapy alone seldom normalizes hyperprolactinaemia within a reasonable time [40]. Combined therapy with bromocriptine is usually necessary to induce ovulation and restore fertility. *Thorner et al.* [89], who used external pituitary irradiation for pretreatment of infertile women with suspected pituitary tumours, found no evidence of tumour expansion during pregnancies in 14 cases. All the women received additional therapy with bromocriptine. *Kelly et al.* [54] advocate pituitary implantation with ^{90}Y as a prophylactic measure in young women with prolactinomas. They found no complications due to tumour expansion during 21 pregnancies in 15 patients with pituitary implants. Both groups concluded that pituitary irradiation is a safe and effective method to limit tumour expansion during pregnancy in women with prolactinomas.

However, irradiation does not seem to completely prevent pituitary tumour complications during pregnancy. Despite prior external pituitary irradiation, 2 patients developed temporary visual field defects [58, 87] and 1 patient had apoplexy of the pituitary tumour with transient paresis of the abducens nerve [59]. *Bergh et al.* [5] described a patient, who despite prior surgery and radiotherapy of a prolactinoma, had further destruction of her pituitary fossa during a gonadotrophin-induced clinically uneventful pregnancy. The long-term effects of radiotherapy on endocrine function in the otherwise healthy young women with infertility due to hyperprolactinaemia are unknown.

After external irradiation treatment for acromegaly, *Eastman et al.* [25] were disappointed to find that there was a definite loss of TSH, ACTH or gonadotrophin function. Moreover, this loss increased with time after radiotherapy. 10 years after radiotherapy, 38% had hypoadrenalism and more than 50% of men and women had hypogonadism. It can thus be questioned whether pituitary irradiation should be used routinely as a prophylactic treatment of young infertile women with prolactinomas until more is known about its long-term effects on pituitary function.

Medical Therapy

Until recently, treatment with bromocriptine was given only as secondary therapy to infertile prolactinoma patients who had previously received surgical treatment or radiotherapy. In the last few years primary treatment with bromocriptine alone has been given to an increasing number of women with infertility due to prolactin-secreting adenomas.

The clinical course and outcome of 31 pregnancies in 26 women with amenorrhoea, hyperprolactinaemia and radiological signs consistent with a pituitary tumour was recently described by *Nillius et al.* [72]. None of the women had received prior tumour therapy either by surgery or irradiation. The asymmetrical and/or enlarged pituitary fossa was classified from the sellar X-rays according to *Thorner et al.* [89]. 17 of the women with suspected prolactinoma had markedly abnormal X-rays (grade B3–B5). Bromocriptine (2.5–20 mg daily, mean 6.5 mg) had been used to induce ovulation in all the women.

Radiological changes of the sella turcica suggestive of pituitary tumour enlargement during pregnancy were found in 4 of the 20 women with term pregnancies on their postpartum X-ray examination. 2 of the 4 women had no symptoms of tumour growth during the pregnancy but there was a slight progression of their radiographic sellar abnormalities (B4→B4E). In 1 of these patients, a 25-year-old woman with a pretreatment prolactin level of 49 μg/l, the sellar erosion had disappeared 1 year after delivery and the pituitary fossa was again classified as B4. No bromocriptine treatment had been given during this period as the patient had regular menstruations and only a slightly raised prolactin level in serum (19 μg/l). A 25-year-old woman with primary amenorrhoea and a pretreatment prolactin level of 75 μg/l complained of severe headache from the second trimester of pregnancy but the visual fields remained normal. The headache disappeared after delivery at term but X-ray revealed that the appearance of the sella turcica had changed from grade B2 to B5E. Bromocriptine therapy (15 mg daily) was reinstituted after weaning because of amenorrhoea and hyperprolactinaemia (75 μg/l). The prolactin levels were restored to normal and menstrual function reappeared. The patient remained normoprolactinaemic after discontinuation of bromocriptine 21 months later, and has continued to have regular menstruations.

Visual field defects were detected at 30 weeks of pregnancy in a 25-year-old woman with 3 years of amenorrhoea and a pretreatment prolactin level of 72 µg/l. The visual impairment improved when bromocriptine therapy was reinstituted and the pregnancy could continue to term. After delivery the visual fields returned to normal but X-ray examination showed destruction of the pituitary fossa with a change from grade B4 to grade B5E (fig. 2). The other 16 women with term pregnancies and the women with spontaneous or induced abortions had no symptoms or signs of tumour enlargement during pregnancy. Many of them had large pituitary tumours (fig. 3, 4).

During the last 2 years, at least 10 reports on bromocriptine treatment of totally 146 infertile women with prolactinomas have appeared in the literature [for references, see 72]. Only 9 complications due to tumour expansion occurred during the 162 pregnancies. Surgical intervention was necessary only in 1 patient, who in the 22nd week of pregnancy developed parasellar expansion of the pituitary tumour with bone destruction and paresis of the right abducens and oculomotor nerves [59]. After transsphenoidal surgery the paresis of both nerves disappeared. The pregnancy continued uneventfully and a normal child was born. The second major complication occurred in the patient described above with visual field defects and destruction of the pituitary fossa during pregnancy (fig. 2). The other 7 cases had only minor temporary visual field defects or radiological signs of slight sellar enlargement postpartum. *Griffith et al.* [41] recently reviewed bromocriptine-induced pregnancies reported to the medical research departments of Sandoz. 9 tumour-related complications occurred during 116 completed pregnancies in mothers with pituitary tumours. Only 2 patients had to undergo surgery. Thus, the frequency of serious complications during pregnancy is low in bromocriptine-treated women with prolactinomas.

A normal X-ray of the sella turcica before pregnancy does not exclude a tumour in the hypothalamic-pituitary region. *Bergh et al.* [4] described a hyperprolactinaemic woman with normal sellar X-ray who, after a bromocriptine-induced clinically uneventful pregnancy, was found to have radiographic evidence of pituitary tumour growth on the postpartum X-ray. We have later seen a similar case with blistering of the sellar floor on the postpartum X-ray. *Burry et al.* [8] described a patient with hyperprolactinaemic amenorrhoea-galactorrhoea and normal sellar X-ray, who in the second month of her bromocriptine-induced pregnancy developed acute visual loss. She was found to have a large suprasellar chromophobe adenoma at craniotomy. Her pregnancy was completed successfully. Recently, *Corenblum* [17]

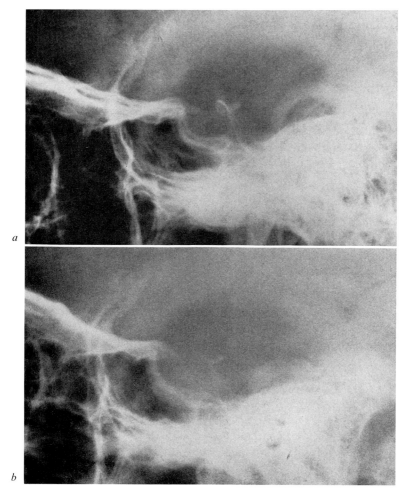

Fig. 2. Sellar X-rays before (a) and during (b) bromocriptine-induced pregnancy in a 25-year-old hyperprolactinaemic woman with amenorrhoea. At 30 weeks of gestation she developed visual field defects and destruction of the pituitary fossa (b).

described another hyperprolactinaemic woman with normal sella turcica tomography who conceived after bromocriptine treatment. At 4 months' gestation she had sudden onset of headaches and was found to have upper quadratic bitemporal defects. A computerized axial tomography (CAT) scan demonstrated a sellar mass with suprasellar extension. Bromocriptine was reinstituted and within 1 week the visual fields were close to normal.

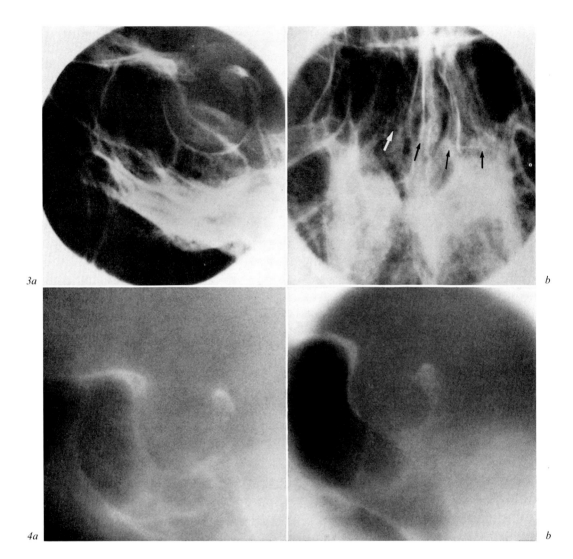

Fig. 3. The markedly asymmetrical and enlarged pituitary fossa of a 31-year-old hyperprolactinaemic woman with amenorrhoea-galactorrhoea who had two uneventful bromocriptine-induced term pregnancies. She was then treated with bromocriptine again. 1 year later X-ray showed evidence of tumour regression (fig. 4).

Fig. 4. Reduction in sellar size after bromocriptine-treatment as indicated by lateral tomograms of the pituitary fossa before (a) and after (b) two uneventful bromocriptine-induced pregnancies followed by bromocriptine treatment in the patient whose pretreatment sellar X-rays are shown in fig. 3. From *Bergh et al.* [unpublished].

A repeat CAT scan failed to demonstrate any sellar mass or suprasellar extension. The last 2 cases illustrate the necessity of excluding suprasellar masses before pregnancy is induced in women with hyperprolactinaemia. The non-invasive technique of computerized axial tomography can nowadays be used to rule out suprasellar extension of pituitary tumours [98]. The above case reports also show that even serious pituitary tumour complications during pregnancy can be managed with a successful outcome.

Management of Pituitary Tumour Complications during Pregnancy

Visual disturbances and other pituitary tumour complications can be managed in several ways during pregnancy. Conservative treatment under close observation is indicated if the symptoms are mild. Medical treatment with bromocriptine in combination with corticosteroids should be given if the symptoms are more pronounced. Bromocriptine has been shown to cause remission of visual field defects and regression of parasellar extension of large prolactinomas (see Medical Treatment of Prolactinomas below). It has also been shown to improve visual field defects during pregnancy [3, 9, 17]. Treatment with corticosteroids in high doses has also proved to normalize visual disturbances during pregnancy [53]. The adrenal steroids may prevent swelling in and around the optic chiasm; they have the additional advantage of depressing adrenal function in mother and child and then the precursors for oestrogen production disappear and oestrogen levels in blood drop dramatically [84]. Finally, treatment with corticosteroids has the advantage that it protects the premature neonate from the respiratory distress syndrome. The effect on the prolactinoma may be similar to that which occurs after delivery of the fetus and placenta. Visual complications due to tumour expansion usually improve rapidly postpartum. If the pregnancy is close to term and the fetus viable, interruption of pregnancy is indicated. If rapid severe deterioration of symptoms occurs emergency surgical decompression should be performed. There are at least 11 reports on acute pituitary surgery during pregnancy with successful outcome for mother and child [for references, see 72].

Medical Treatment of Prolactinomas

In animal studies, ergot alkaloid derivatives were found to induce regression or inhibition of pituitary tumour growth. Ergocornine caused a decrease in cells and a disappearance or pycnosis of nuclei in rat pituitary tumour tissue [77]. *MacLeod and Lehmeyer* [63] found that ergotamine, ergocryptine and ergocornine were all effective in suppressing growth and function of several hormone-producing pituitary tumours in rats by a direct effect on the tumour. It was suggested that the suppression of the tumour could be the result of vasoconstrictive effects of the ergot alkaloids rather than due to their dopaminergic properties. Bromocriptine has less potent vasopressor effects than ergotamine and ergocryptine [32]. In a recent study, *Lamberts and MacLeod* [57] unexpectedly found that chronic administration of high doses of bromocriptine failed to produce a suppressive effect on tumour growth in rats bearing transplanted prolactin-secreting pituitary tumours. However, bromocriptine has been shown to inhibit proliferation of prolactin cells in the rat by reducing oestrogen-induced stimulation of DNA synthesis and pituitary mitotic activity [20, 62, 74, 86]. During recent years, several clinical investigators have described anti-tumour effects of bromocriptine in patients with prolactin-secreting pituitary tumours.

Improvement of Visual Field Defects during Bromocriptine Treatment
Case reports, which briefly described improvement of visual field defects during bromocriptine treatment of patients with large prolactinomas, suggested that bromocriptine may cause a regression of pituitary tumours [16, 36, 44, 49, 95]. In 2 patients with acromegaly, *Wass et al.* [94] observed disappearance of bilateral visual field defects for red vision within 3 months of starting treatment with bromocriptine. *Vaidya et al.* [92] described in detail normalization of bilateral hemianopsia after 2 months of bromocriptine treatment of 2 women with hyperprolactinaemic amenorrhoea. Bromocriptine treatment has also been shown to improve or normalize visual impairment due to pituitary tumour expansion during pregnancy [3, 9, 17]. Thus, indirect clinical evidence from improvement of visual field defects suggested that bromocriptine can reduce the size of prolactin-secreting pituitary tumours.

Regression of Prolactinomas during Long-Term Treatment with Bromocriptine
Findings during transsphenoidal surgical exploration of the pituitary fossa suggested that tumour regression had occurred during long-term treatment with bromocriptine in 1 of our patients with clear clinical, biochemical and radiological evidence of a prolactin-secreting pituitary adenoma [71]. She was a nulliparous 27-year-old woman with 12 years of amenorrhoea-galactorrhoea, hyperprolactinaemia (620 µg/l) and an asymmetrically en-

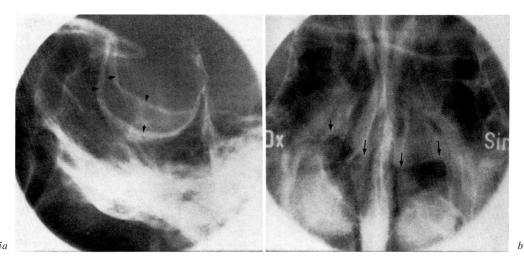

Fig. 5. The pretreatment sellar X-ray of the 32-year-old hyperprolactinaemic woman with 13 years of amenorrhoea-galactorrhoea, in whom bromocriptine therapy possibly induced regression of the pituitary tumour. For further details see text.

larged sella turcica (fig. 5). During bromocriptine treatment the raised prolactin levels normalized and ovulatory menstruations were regained after 48 weeks of treatment (fig. 6). At transsphenoidal surgery after 27 months of treatment, no tumour was found. Bromocriptine was discontinued, the patient continued to ovulate and conceived spontaneously. During the uneventful pregnancy the serum prolactin levels slowly increased to reach a maximum of 78 µg/l at term. After delivery the prolactin levels decreased but levelled at around 50 µg/l. Menstrual cycles returned but during the first year after delivery they have been slightly irregular. Progesterone determinations indicate inadequate luteal function.

Another amenorrhoeic patient with hyperprolactinaemia (290 µg/l) and a grossly asymmetrical and enlarged sella turcica remained normoprolactinaemic with normal ovulatory cycles after discontinuation of 21 months of bromocriptine therapy. Pretreatment air encephalography had shown slight suprasellar extension of a pituitary tumour [6]. Bromocriptine therapy was discontinued after 12–30 months of treatment in another 10 hyperprolactinaemic women with radiological signs consistent with a pituitary tumour. However, in all of these women, hyperprolactinaemia and amenorrhoea or anovulatory cycles returned after discontinuation of treatment [7]. Thus, in

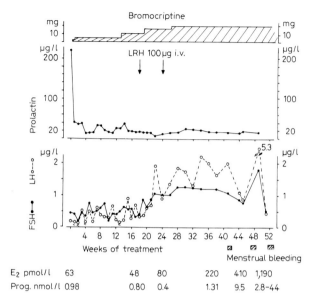

Fig. 6. Basal levels of prolactin, FSH and LH before and during the first year of bromocriptine treatment of the woman whose sellar X-rays are shown in figure 5.

the majority of the prolactinoma patients, long-term treatment with bromocriptine was not associated with a permanent normalization of the prolactin hypersecretion.

Persisting suppression of prolactin secretion after long-term treatment with bromocriptine was recently described by *Eversmann et al.* [28] in 10 patients with macroprolactinomas and excessively high postoperative prolactin levels. After withdrawal of bromocriptine the prolactin levels increased to only 36% of the pretreatment level. This effect was not observed in patients with prolactinomas with only moderately raised serum prolactin concentrations. The authors interpreted the persisting suppression of the prolactin secretion as evidence of shrinkage of the large prolactinomas.

Radiographic Evidence of Reduction of Pituitary Tumour Size During Treatment with Dopamine Agonists

During the last few years, shrinkage of prolactinomas has been reported to occur in association with dopamine agonist therapy in several studies, in which objective radiological evidence confirmed clinical and biochemical signs of tumour regression. Canadian investigators [15, 29, 38] described shrinkage of large prolactinomas with suprasellar extension and visual field defects during bromocriptine treatment. The visual fields were restored to normal and repeat pneumoencephalograms showed disappearance of the suprasellar extension. Partially empty sellae with air entering the sella were observed in 3 cases after bromocriptine therapy for at least one year [29, 38]. *Sobrinho et al.* [85] described partial remission of a prolactinoma, which had grown invasively into the sphenoidal sinus. Air contrast tomograms showed reduction of the soft tissue invasion after 13 months of bromocriptine treatment.

A reduction of pituitary fossa size was found by *Wass et al.* [94] to occur after long-term bromocriptine treatment in about 20% of 69 patients with prolactinomas or acromegaly. 5 of 26 prolactinoma patients had evidence of reduced fossa size after treatment with bromocriptine for 19–78 months. 3 of these 5 patients had also had external pituitary irradiation. Shrinkage of the pituitary tumour was observed in 8 of 43 bromocriptine-treated patients with acromegaly. *Mühlenstedt et al.* [68] described a patient with a macroadenoma, which expanded during a bromocriptine-induced pregnancy. Postpartum, bromocriptine treatment was reinstituted. Radiological signs of tumour regression, i.e. recalcification of dorsum sellae and posterior clinoids and decrease of sellar volume, were observed already after 4 months of treatment.

More recently, reduction in size of pituitary tumours during dopamine agonist therapy has been objectively demonstrated by repeat CAT. *McGregor et al.* [66] demonstrated abolition of the extrasellar extension of a locally invasive prolactinoma both by clinical assessment and by CAT after treatment with bromocriptine (fig. 7, 8). Their male patient had previously been treated with both surgery and irradiation but the tumour recurred. Similar cases were reported by *Matsumura et al.* [65] and by *Landolt et al.* [60]. The possibility of a synergism between bromocriptine and radiotherapy was discussed by *Landolt et al.* [60] who suggested combined treatment with radiotherapy and bromocriptine in patients with invasive prolactinomas. However, complete disappearance of suprasellar and parasellar extensions of prolactinomas have also been described during dopamine agonist therapy in patients not previously treated with radiotherapy [14, 95]. *Chiodini et al.* [14] treated a patient with a large prolactinoma with the dopamine agonist lisuride (1.6 mg/day). After 1 year of treatment, repeat CAT scans showed a marked reduction of the suprasellar portion of the tumour.

A recent prospective study [67] on the effects of bromocriptine on pituitary tumour size is of great interest. 12 pituitary tumour patients were

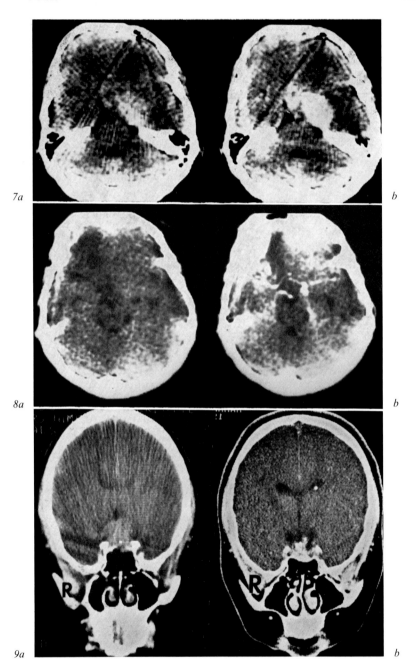

treated for 3 months with 20 mg of bromocriptine daily. The effects of the treatment on tumour size were assessed by computerized tomography and metrizamide cisternography [43]. All the 5 patients (4 men) with prolactinoma showed a rapid and dramatic improvement. There was radiographic evidence of reduction in the size of the pituitary tumour already after 3 months of treatment. This was accompanied by a fall in prolactin concentrations and clinical and biochemical improvement in their hypopituitarism. A visual field defect in 1 of the patients resolved. 3 of these patients had previously been unsuccessfully treated with both transfrontal craniotomy and irradiation. The authors stressed the potential value of bromocriptine in treating recurrences after surgery or radiotherapy. They suggested that bromocriptine should be used as primary treatment for large prolactinomas.

Regression of pituitary prolactinomas during bromocriptine treatment can be extremely rapid. *Thorner et al.* [91] recently reported on results of primary treatment with bromocriptine in 2 men with large prolactin-secreting pituitary tumours. Headache, visual acuity and visual fields improved after only 3 days of treatment in 1 of the patients who had marked visual impairment. Already after 2 weeks of treatment, a repeat CAT scan showed marked reduction in tumour size (fig. 9). In another male patient, metrizamide cisternography with tomography demonstrated definite tumour regression with a partially empty fossa due to shrinkage of the pituitary mass after 6 weeks of treatment. It is interesting to note that these remarkable rapid results were obtained with the small daily dose of 7.5 mg of bromocriptine [91].

Fig. 7. CAT scans before (a) and after (b) i.v. contrast medium in a male with hyperprolactinaemia. The prolactinoma can be seen extending above and laterally to the sella turcica. From *McGregor et al.* [66], with permission.

Fig. 8. CAT scans before (a) and after (b) i.v. contrast medium in the patient described in figure 7 after a 3-month course of bromocriptine (5 mg q.d.s.). No evidence of tumour can be found. From *McGregor et al.* [66], with permission.

Fig. 9. Coronal CAT head scans before and after 2 weeks of bromocriptine therapy, illustrating rapid regression of a prolactin-secreting pituitary adenoma. Before therapy scan (a) shows a large enhancing mass in the pituitary fossa extending inferiorly into sphenoidal sinus and superiorly into chiasmatic cistern and abutting into third ventricle. 2 weeks after starting bromocriptine, therapy scan (b) shows marked reduction in tumor size with regression of the suprasellar extension. From *Thorner et al.* [91], with permission.

Conclusions

Review of the literature indicates that neither surgery nor irradiation reliably provides the definitive cure that had been hoped for the patients with prolactin-secreting pituitary tumours. In the last few years, experience with medical treatment has revealed that induction of ovulation with bromocriptine is remarkably safe both in patients with microtumours and those with the smaller macrotumours. In the future it is possible that even the larger macrotumours and these with suprasellar extension can be managed by medical therapy with dopamine agonists.

References

1 Antunes, J.L.; Housepian, E.M.; Frantz, A.G.; Holub, D.A.; Hui, R.M.; Carmel, P.W., and Quest, D.O.: Prolactin-secreting pituitary tumors. Ann. Neurol. 2: 148–153 (1977).
2 Bergh, T.; Nillius, S.J., and Wide, L.: Hyperprolactinaemic amenorrhoea – results of treatment with bromocriptine. Acta endocr., Copenh. 88: suppl. 216, 147–164 (1978).
3 Bergh, T.; Nillius, S.J., and Wide, L.: Clinical course and outcome of pregnancies in amenorrhoeic women with hyperprolactinaemia and pituitary tumours. Br. med. J. i: 875–880 (1978).
4 Bergh, T.; Nillius, S.J., and Wide, L.: Bromocriptine treatment of 42 hyperprolactinaemic women with secondary amenorrhoea. Acta endocr., Copenh. 88: 435–451 (1978).
5 Bergh, T.; Nillius, S.J., and Wide, L.: Bromocriptine treatment of seven women with primary amenorrhoea and prolactin-secreting pituitary tumours. Clin. Endocrinol. 10: 145–154 (1979).
6 Bergh, T.; Nillius, S.J.; Lundberg, P.O., and Wide, L.: Bromocriptine treatment of prolactinomas. New Engl. J. Med. 300: 1391 (1979).
7 Bergh, T.; Nillius, S.J., and Wide, L.: Long-term bromocriptine therapy of women with prolactin-secreting pituitary tumours. 6th Int. Congr. of Endocrinology, Melbourne 1980, abstr. 859.
8 Burry, K.A.; Schiller, H.S.; Mills, R.; Harris, B., and Heinrichs, L.: Acute visual loss during pregnancy after bromocriptine-induced ovulation. Obstet. Gynec., N.Y. 52: 19s–22s (1978).
9 Burzaco, I.; Gonzáles-Merlo, J., and Cámara, J. de la: Bromocriptina en la expansión de tumores hipofisarios durante el embarazo. Clin. Invest. Ginec. Obstet. 5: 90–92 (1978).
10 Buvat, J.; Asfour, M.; Buvat-Herbaut, M., and Fossati, P.: Prolactin and human sexual behaviour; in Robyn and Harter, Progress in prolactin physiology and pathology, pp. 317–329 (Elsevier, North-Holland 1977).
11 Carter, J.N.; Tyson, J.E.; Tolis, G.; Van Vliet, S.; Faiman, C., and Friesen, H.G.:

Prolactin-secreting tumors and hypogonadism in 22 men. New Engl. J. Med. 299: 847–852 (1978).

12 Child, D.F.; Gordon, H.; Mashiter, K., and Joplin, G.F.: Pregnancy, prolactin, and pituitary tumours. Br. med. J. iv: 87–89 (1975).

13 Chiodini, P.G.; Liuzzi, A.; Muller, E.E.; Botalla, L.; Cremascoli, G.; Oppizzi, G.; Verde, G., and Silvestrini, F.: Inhibitory effect of an ergoline derivative, methergoline, on growth hormone and prolactin levels in acromegalic patients. J. clin. Endocr. Metab. 43: 356 (1976).

14 Chiodini, P.G.; Liuzzi, A.; Silvestrini, F.; Verde, G.; Cozzi, R.; Marsili, M.T.; Horowski, R.; Passerini, F.; Luccarelli, G., and Borghi, P.G.: Size reduction of a prolactin secreting adenoma during a long-term treatment with a dopamine-agonist lisuride. Clin. Endocrinol. 12: 47–51 (1980).

15 Corenblum, B.: Bromocriptine in pituitary tumours. Lancet ii: 786 (1978).

16 Corenblum, B.; Webster, B.R.; Mortimer, C.B., and Ezrin, C.: Possible anti-tumour effect of 2 bromoergocryptine (CB-154 Sandoz) in 2 patients with large prolactin-secreting pituitary adenomas. Clin. Res. 23: 614A (1975).

17 Corenblum, B.: Successful outcome of ergocryptine-induced pregnancies in twentyone women with prolactin-secreting pituitary adenomas. Fert. Steril. 32: 183–186 (1979).

18 Corrodi, H.; Fuxe, K.; Hökfelt, T.; Tidbrink, P., and Ungerstedt, U.: Effect of ergot drugs on central catecholamine neurons. Evidence for a stimulation of central dopamine neurons. J. Pharm. Pharmac. 25: 409 (1973).

19 Crosignani, P.G.; Ferrari, C.; Mattei, A.; Fadini, R.; Meschia, M.; Caldara, R.; Rampini, P.; Telloni, P., and Reschini, E.: Metergoline treatment of hyperprolactinemic states. Fert. Steril. 32: 280–285 (1979).

20 Davies, C.; Jacobi, J.; Lloyd, H.M., and Meares, J.D.: DNA synthesis and the secretion of prolactin and growth hormone by the pituitary gland of the male rat: effects of diethylstilboestrol and 2-bromo-α-ergocryptine methanesulphonate. J. Endocr. 61: 411–417 (1974).

21 DeCecco, L.; Foglia, G.; Ragni, N.; Rossato, P., and Venturini, P.L.: The effect of lisuride hydrogen maleate in the hyperprolactinaemia-amenorrhoea syndrome: clinical and hormonal responses. Clin. Endocrinol. 9: 491–498 (1978).

22 Delitala, G.; Wass, J.A.H.; Stubbs, W.A.; Jones, A.; Williams, S., and Besser, G.M.: The effect of lisuride hydrogen maleate, an ergot derivative on anterior pituitary hormone secretion in man. Clin. Endocrinol. 11: 1–9 (1979).

23 Dericks-Tan, J.S.E.; Bradler, H., and Taubert, H.-D: Evaluation of proteohormon concentrations in serum and seminal fluid and its relationship with sperm count. Andrologia 10: 175–182 (1978).

24 Derome, P.J.; Peillon, F.; Bard, R.H.; Jedynak, C.P.; Racadot, J. et Guiot, G.: Adénomes à prolactine: résultats du traitement chirurgical. Nouv. Presse méd. 8: 577–583 (1979).

25 Eastman, R.; Gorden, P., and Roth, J.: Conventional supervoltage irradiation is an effective treatment for acromegaly. J. clin. Endocr. Metab. 48: 931–940 (1979).

26 Enoksson, P.; Lundberg, N.; Sjöstedt, S., and Skanse, B.: Influence of pregnancy on visual fields in suprasellar tumours. Acta pychiat. neurol. scand. 36: 524–538 (1961).

27 Erdheim, J. und Stumme, E.: Über die Schwangerschaftsveränderung der Hypophyse; Beiträge zur pathologischen Anatomie und zur allgemeinen Pathologie, vol. 46, pp. 1–132 (Fischer, Jena 1909).

28 Eversmann, T.; Fahlbusch, R.; Rjosk, H.K., and Werder, K. von: Persisting suppression of prolactin secretion after long-term treatment with bromocriptine in patients with prolactinomas. Acta endocr., Copenh. *92:* 413–427 (1979).
29 Ezrin, C.; Kovacs, K., and Horvath, E.: Hyperprolactinemia. Morphologic and clinical considerations. Med. Clins. N. Am. *62:* 393–408 (1978).
30 Fahlbusch, R.: Surgical failures in prolactinomas. Proc. 2nd European Workshop on Pituitary Adenomas, Paris 1979 (in press).
31 Ferrari, C.; Caldara, R.; Rampini, P.; Telloli, P.; Romussi, M.; Bertazzoni, A.; Polloni, G.; Mattei, A., and Crosignani, P.G.: Inhibition of prolactin release by serotonin antagonists in hyperprolactinemic subjects. Metabolism *27:* 1499–1504 (1978).
32 Fluckiger, E.: The pharmacology of bromocriptine; in Bayliss, Turner and Maclay, Pharmacological and clinical aspects of bromocriptine (Parlodel), pp. 12–26 (MCS Consultants, Tunbridge Wells 1976).
33 Fluckiger, E.: Effects of bromocriptine on the hypothalamopituitary axis. Acta endocr., Copenh. *88:* suppl. 216, pp. 111–117 (1978).
34 Franks, S.; Jacobs, H.S.; Hull, M.G.R.; Steele, S.J., and Nabarro, J.D.N.: Management of hyperprolactinaemic amenorrhoea. Br. J. Obstet. Gynaec. *84:* 241–253 (1977).
35 Franks, S.; Jacobs, H.S.; Martin, N., and Nabarro, J.D.N.: Hyperprolactinaemia and impotence. Clin. Endocrinol. *8:* 277–287 (1978).
36 Friesen, H.G. and Tolis, G.: The use of bromocriptine in the galactorrhoea-amenorrhoea syndromes: The Canadian cooperative study. Clin. Endocrinol. *6:* 91s–99s (1977).
37 Gemzell, C. and Wang, C.F.: Outcome of pregnancy in women with pituitary adenoma. Fert. Steril. *31:* 363–372 (1979).
38 George, S.R.; Burrow, G.N.; Zinman, B., and Ezrin, C.: Regression of pituitary tumors, a possible effect of bromoergocryptine. Am. J. Med. *66:* 697–702 (1979).
39 Goluboff, L.G. and Ezrin, C.: Effect of pregnancy on the somatotroph and the prolactin cell of the human adenohypophysis. J. clin. Endocr. Metab. *29:* 1533–1538 (1969).
40 Gómez, F.; Reyes, F.I., and Faiman, C.: Nonpuerperal galactorrhea and hyperprolactinemia. Clinical findings, endocrine features and therapeutic responses in 56 cases. Am. J. Med. *62:* 648–660 (1977).
41 Griffith, R.W.; Turkalj, I., and Braun, P.: Pituitary tumours during pregnancy in mothers treated with bromocriptine. Br. J. clin. Pharmacol. *7:* 393–396 (1979).
42 Guiot, G.: Transsphenoidal approach in surgical treatment of pituitary adenomas; in Kohler and Ross, Diagnosis and treatment of pituitary tumours, pp. 158–178 (Excerpta Medica, Amsterdam 1973).
43 Hall, K. and McAllister, V.L.: Metrizamide cisternography in pituitary and juxtapituitary lesions. Radiology *134:* 101–108 (1980).
44 Hancock, K.W.; Scott, J.S.; Gibson, R.M., and Lamb, J.T.: Pituitary tumours and pregnancy, Br. med. J. *i:* 1487–1488 (1978).
45 Hardy, J.: Transsphenoidal surgery of hypersecreting pituitary tumours; in Kohler and Ross, Diagnosis and treatment of pituitary tumors, pp. 179–194 (Excerpta Medica, Amsterdam 1973).
46 Hardy, J.; Beauregard, H., and Robert, F.: Prolactin-secreting pituitary adenomas:

transsphenoidal microsurgical treatment; in Robyn and Harter, Progress in prolactin physiology and pathology, pp. 361–370 (Elsevier/North-Holland, Amsterdam 1978).

47 Hargreave, T.B.; Kyle, K.F.; Kelly, A.M., and England, P.: Prolactin and gonadotrophins in 208 men presenting with infertility. Br. J. Urol. 49: 747–750 (1977).

48 Horowski, R.; Wendt, H., and Gräf, K.-J.: Prolactin-lowering effect of low doses of lisuride in man. Acta endocr., Copenh. 87: 234–240 (1978).

49 Huth, J. and Tyson, J.E.: Incomplete suppression of hyperprolactinemic galactorrhea-amenorrhea with bromoergocryptine. Clin. Res. 25: 295A (1977).

50 Hökfelt, T. and Fuxe, K.: On the morphology and the neuroendocrine role of the hypothalamic catecholamine neurons; in Knigge, Scott and Weindl, Brain-endocrine interaction. Median eminence: structure and function, pp. 181 (Karger, Basel 1972).

51 Hökfelt, B. and Nillius, S.J.: The dopamine agonist bromocriptine. Theoretical and clinical aspects. Acta endocr., Copenh. 216: suppl., pp. 1–230 (1978).

52 Jéquier, A.M.; Crich, J.C., and Ansell, I.D.: Clinical findings and testicular histology in three hyperprolactinemic infertile men. Fert. Steril. 31: 525–530 (1977).

53 Jewelewicz, R.; Zimmerman, E.A., and Carmel, P.W.: Conservative management of a pituitary tumor during pregnancy following induction of ovulation with gonadotropins. Fert. Steril. 28: 35–40 (1977).

54 Kelly, W.F.; Doyle, F.H.; Mashiter, K.; Banks, L.M.; Gordon, H., and Joplin, G.F.: Pregnancies in women with hyperprolactinaemia: clinical course and obstetric complications of 41 pregnancies in 27 women. Br. J. Obstet. Gynaec. 86: 698–705 (1979).

55 Keye, W.R.; Chang, R.J.; Monroe, S.E.; Wilson, C.B., and Jaffe, R.B.: Prolactin-secreting pituitary adenomas in women. II. Menstrual function, pituitary reserves, and prolactin production following microsurgical removal. Am. J. Obstet. Gynec. 134: 360–365 (1979).

56 Kleinberg, D.L.; Noel, G.L., and Frantz, A.G.: Galactorrhea: a study of 235 cases, including 48 with pituitary tumors. New Engl. J. Med. 296: 589–600 (1977).

57 Lamberts, S.W.J. and MacLeod, R.M.: The inability of bromocriptine to inhibit prolactin secretion by transplantable rat pituitary tumors: observations on the mechanism and dynamics of the autofeedback regulation of prolactin secretion. Endocrinology 104: 65–70 (1979).

58 Lamberts, S.W.J.; Seldenrath, H.J.; Kwa, H.G., and Birkenhäger, J.C.: Transient bitemporal hemianopsia during pregnancy after treatment of galactorrhea-amenorrhea syndrome with bromocriptine. J. clin. Endocr. Metab. 44: 180–184 (1977).

59 Lamberts, S.W.J.; Klijn, J.G.M.; DeLange, S.A.; Singh, R.; Stefanko, S.Z., and Birkenhäger, J.C.: The incidence of complications during pregnancy after treatment of hyperprolactinemia with radiologically evident pituitary tumors. Fert. Steril. 31: 614–619 (1979).

60 Landolt, A.M.; Wuthrich, R., and Fellmann, H.: Regression of pituitary prolactinoma after treatment with bromocriptine. Lancet i: 1082–1083 (1979).

61 Liuzzi, A.; Chiodini, P.G.; Oppizzi, G.; Botalla, L.; Verde, G.; De Stefano, L.; Colussi, G.; Gräf, K.J., and Horowski, R.: Lisuride hydrogen maleate: evidence for a long lasting dopaminergic activity in human. J. clin. Endocr. Metab. 46: 196–202 (1978).

62 Lloyd, H.M.; Meares, J.D., and Jacobi, J.: Effects of oestrogen and bromocryptine on *in vivo* secretion and mitosis in prolactin cells. Nature, Lond. 255: 497–498 (1975).

63   MacLeod, R.M. and Lehmeyer, J.E.: Suppression of pituitary tumor growth and function by ergot alkaloids. Cancer Res. *33:* 849–855 (1973).
64   Magyar, D.M. and Marshall, J.R.: Pituitary tumors and pregnancy. Am. J. Obstet. Gynec. *132:* 739–751 (1978).
65   Matsumura, S.; Mori, S., and Uozumi, T.: Size reduction of a large prolactinoma by bromoergocryptine (CB-154) treatment. 2nd European Workshop on Pituitary Adenomas, Paris 1979, abstr. 84.
66   McGregor, A.M.; Scanlon, M.F.; Hall, K.; Cook, D.B., and Hall, R.: Reduction in size of a pituitary tumor by bromocriptine therapy. New Engl. J. Med. *300:* 291–293 (1979).
67   McGregor, A.M.; Scanlon, M.F.; Hall, R., and Hall, K.: Effects of bromocriptine on pituitary tumour size. Br. med. J. *ii:* 700–703 (1979).
68   Mühlenstedt, D.; Osmers, F. und Schneider, H.P.G.: Regression eines Hypophysenadenoms unter Bromocriptin. Arch. Gynaec. *226:* 341–346 (1978).
69   Nabarro, J.D.N. and Franks, S.: Prolactin and amenorrhoea; in Jacobs, Advances in gynaecological endocrinology. Proc. 6th Study Group of the Royal College of Obstetricians and Gynaecologists, pp. 248–261 (London 1979).
70   Nagulesparen, M.; Ang, V., and Jenkins, J.S.: Bromocriptine treatment of males with pituitary tumours, hyperprolactinaemia, and hypogonadism. Clin. Endocrinol. *9:* 73–79 (1978).
71   Nillius, S.J.; Bergh, T.; Lundberg, P.O.; Stahle, J., and Wide, L.: Regression of a prolactin-secreting pituitary tumor during long-term treatment with bromocriptine. Fert. Steril. *30:* 710–712 (1978).
72   Nillius, S.J.; Bergh, T., and Larsson, S.-G.: Pituitary tumours and pregnancy. Proc. 2nd European Workshop on Pituitary Adenomas, Paris 1979, in press.
73   Pasteels, J.L.: Recherches morphologiques et expérimentales sur la sécrétion de prolactine. Archs Biol., Liège *74:* 439 (1963).
74   Pawlikowski, M.; Kunert-Radek, J., and Stepien, H.: Direct antiproliferative effect of dopamine agonists on the anterior pituitary gland in organ culture. J. Endocrinol. *79:* 245–246 (1978).
75   Pepperell, R.J.; McBain, J.C., and Healy, D.L.: Ovulation induction with bromocriptine (CB 154) in patients with hyperprolactinaemia. Aust. N.Z. J. Obstet. Gynaec. *17:* 181–191 (1977).
76   Post, K.D.; Biller, B.J.; Adelman, L.S.; Molitch, M.E.; Wolpert, S.M., and Reichlin, S.: Selective transphenoidal adenomectomy in women with galactorrhea-amenorrhea. J. Am. med. Ass. *242:* 158–162 (1979).
77   Quadri, S.K.; Lu, K.H., and Meites, J.: Ergot-induced inhibition of pituitary tumor growth in rats. Science *176:* 417–418 (1972).
78   Reuss, A. von: Sehnervenleiden in Folge von Gravidität. Wien. klin. Wschr. *31:* 1116–1119 (1908).
79   Rjosk, H.-K. and Schill, W.-B.: Serum prolactin in male infertility. Andrologia *11:* 297–304 (1979).
80   Rjosk, H.-K.; Fahlbusch, R. und Werder, K. von: Hyperprolaktinämie und Sterilität. Arch. Gynaec. *228:* 518–530 (1979).
81   Roulier, R.; Mattei, A., and Franchimont, P.: Prolactin in male reproductive functions; in Robyn and Harter, Progress in prolactin physiology and pathology, pp. 305–316 (Elsevier/North-Holland, Amsterdam 1978).

82 Segal, S.; Polishuk, W.Z., and Ben-David, M.: Hyperprolactinemic male infertility. Fert. Steril. *27:* 1425–1427 (1976).
83 Shearman, R.P. and Fraser, I.S.: Impact of new diagnostic methods on the differential diagnosis and treatment of secondary amenorrhoea. Lancet *i:* 1195–1197 (1977).
84 Simmer, H.H.; Tulchinsky, C.; Gold, E.M.; Frankland, M.; Greipel, M., and Gold, A.S.: On the regulation of estrogen production by cortisol and ACTH in human pregnancy at term. Am. J. Obstet. Gynec. *119:* 283–296 (1974).
85 Sobrinho, L.G.; Nunes, M.C.P.; Santos, M.A., and Mauricio, J.C.: Radiological evidence for regression of prolactinoma after treatment with bromocriptine. Lancet *ii:* 257–258 (1978).
86 Stepien, H.; Wolaniuk, A., and Pawlikowski, M.: Effects of pimozide and bromocriptine on anterior pituitary cell proliferation. J. neural. Transm. *42:* 239–244 (1978).
87 Thorner, M.O.; Besser, G.M.; Jones, A.; Dacie, J., and Jones, A.E.: Bromocriptine treatment of female infertility: report of 13 pregnancies. Br. med. J. *iv:* 694–697 (1975).
88 Thorner, M.O. and Besser, G.M.: Bromocriptine treatment of hyperprolactinaemic hypogonadism. Acta endocr., Copenh. *88:* suppl. 216, pp. 131–146 (1978).
89 Thorner, M.O.; Edwards, C.R.W.; Charlesworth, M.; Dacie, J.E.; Moult, P.J.A.; Rees, L.H.; Jones, A.E., and Besser, G.M.: Pregnancy in patients presenting with hyperprolactinaemia. Br. med. J. *ii:* 771–774 (1979).
90 Thorner, M.O.; Evans, W.S.; MacLeod, R.M.; Nunley, W.C., Jr.; Rogol. A.D.; Morris, J.L., and Besser, G.M.: Hyperprolactinemia: current concepts of management including medical therapy with bromocriptine; in Goldstein, Calne, Lieberman and Thorner, Ergot compounds and brain function – neuroendocrine and neuropsychiatric aspects pp. 165–190 (Raven Press, New York, 1980).
91 Thorner, M.O.; Martin, W.H.; Rogol, A.D.; Morris, J.L.; Perryman, R.L.; Conway, B.P.; Howards, S.S.; Wolfman, M.G., and MacLeod, R.M.: Rapid regression of pituitary prolactinomas during bromocriptine treatment. J. clin. Endocr. Metab. (in press, 1980).
92 Vaidya, R.A.; Aloorkar, S.D.; Rege, N.R.; Maskati, B.T.; Jahangir, R.P.; Sheth, A.R., and Pandya, S.K.: Normalization of visual fields following bromocriptine treatment in hyperprolactinemic patients with visual field constriction. Fert. Steril. *29:* 632–636 (1978).
93 Wass, J.A.H.; Thorner, M.O.; Morris, D.V.; Rees, L.H.; Mason, A.S.; Jones, A.E., and Besser, G.M.: Long-term treatment of acromegaly with bromocriptine. Br. med. J. *i:* 875–878 (1977).
94 Wass, J.A.H.; Moult, P.J.A.; Thorner, M.O.; Dacie, J.E.; Charlesworth, M.; Jones, A.E., and Besser, G.M.: Reduction of pituitary-tumour size in patients with prolactinomas and acromegaly treated with bromocriptine with or without radiotherapy. Lancet *ii:* 66–69 (1979).
95 Werder, K. von; Fahlbusch, R.; Landgraf, R.; Pickardt, C.R.; Rjosk, H.-K., and Scriba, P.C.: Treatment of patients with prolactinomas. J. Endocr. Invest. *1:* 47–58 (1978).
96 Werder, K. von; Eversmann, T.; Fahlbusch, R., and Rjosk, H.-K.: Bromocriptine treatment of prolactinomas. New Engl. J. Med. *300* 1391–1392 (1979).
97 Wiebe, R.H.; Kramer, R.S., and Hammond, C.B.: Surgical treatment of prolactin-secreting microadenomas. Am. J. Obstet. Gynec. *134:* 49–55 (1979).

98 Wolpert, S.M.; Post, K.D.; Biller, B.J., and Molitch, M.E.: The value of computed tomography in evaluating patients with prolactinomas. Radiology *131:* 117–119 (1979).

S.J. Nillius, MD, PhD, Department of Obstetrics and Gynaecology, University Hospital, S-750 14 Uppsala (Sweden)

Discussion

Friesen: Are monthly visual-field examinations and a postoperative X-ray (particularly the repeat X-rays) really mandatory, given the low complication rate, knowing the long natural history of this syndrome?

Nillius: I think that the visual field examinations are really necessary to see the complications as early as possible. With regard to the X-ray, I tend to agree with you that perhaps we could leave it. We do not do tomographs, we do ordinary X-rays. If there are no gross changes, we do not do repeat tomograms, that gives a patient too much radiation. At this stage, it is very important to document whether we can see any changes, to obtain estimates of the incidence of these complications. When we have enough material, we may be able to change the recommendation, but at the moment, it is advisable to do an ordinary X-ray at least.

Friesen: Is it your impression that measuring serum prolactin is a reasonably good guide to tumour size in the follow-up of patients? As good, perhaps, as X-rays?

Nillius: I do not think that, during pregnancy, there is much value in prolactin determinations. When you discontinue bromocriptine therapy, the patients return to their pretreatment prolactin levels and stay there with very few changes during pregnancy. It has been suggested that you should do monthly prolactin values during pregnancy. I cannot see that it would change our management of the patient in any way if the prolactin level was a little high and the patient has a normal visual field and no other complications.

Thorner: Most people would now agree that the data are not there to support the view that the slowing of dopamine turnover in the hypothalamus was a reflection of the stimulation of receptors of dopamine in the hypothalamus by bromocriptine but that it is due to the fall in prolactin, and it was the fall in prolactin that led to the slowing in turnover. Another comment concerns the case reported by *Lamberts* in 1977. If you read that paper very carefully, there is actually no data on the pre-pregnancy state of the pituitary. The pneumo-encephalogram was not performed prior to the pregnancy, and minor field deficiencies were present before the pregnancy ever started. Although I agree with you that there may, occasionally, be a patient who develops complications following irradiation, there are very few documented cases: many of those that are quoted are inadequately documented before.

Nillius: I thought that you would comment about your case, which maybe questionable too.

Frantz: In the 26 patients with untreated prolactinomas that you reported, who were given bromocriptine, how many were micro-adenomas and how many were macroadenomas?

Nillius: That is difficult to answer, because the definition of a micro-adenoma is the surgical size of the tumour, which we do not know. That raises the problems we have in classifying the tumour. We should try to find a good classification by which those who do not operate could classify their patients.

Dolman: The incidence of teratogenicity in patients treated with bromocriptine is very low. However, looking at the data more carefully, very few of the patients had been treated for much longer than 5 weeks, and many of them had been treated for only 14 days. There is now substantial evidence that there is some shrinkage, resolution, evolution, whatever you want to call it, of these tumours. The question, therefore, is whether such patients should be treated during the entire pregnancy with bromocriptine. I have talked to several groups who claim that they have done this and had no problems, but I have yet to see the data. I was wondering what your opinion was of this type of approach; it might prevent some of the complications.

Nillius: You are absolutely right but we do not favour it at the moment. The incidence of complications is so low and they can be managed adequately. I would hesitate to use bromocriptine prophylactically throughout pregnancy.

Carter: Is there any evidence that the response of tumours to bromocriptine is different if the tumours secrete excessively both growth hormone and prolactin?

Thorner: The experience, which they have had at St. Bartholomew's at least, is that 2 or 3 of the patients who had combined hyperprolactinaemia and acromegaly showed regression, or changes in the pituitary fossa size, or changes on CT scan. No patient with acromegaly alone, with normal prolactin levels, showed a reduction of tumour size. So, I think that probably the hyperprolactinaemia is the important marker, but if there is also growth hormone hypersecretion, that does not mean that the patient will not respond.

Cowden: You were very careful to point out the difference in incidence of complications after bromocriptine-induced pregnancies in patients with small, as compared with patients with large, prolactinomas. It is also important to emphasize, in reviewing the surgical data, that the vast majority of these series were made up of patients who had very large tumours. It is, therefore, not surprising that the clinical and biochemical cure rate was less than one might have desired. Similarly, the incidence of panhypopituitarism post-operatively was considerable. It is now quite clear that, if these tumours are treated when they are small, the results, both clinical and biochemical and in terms of preservation of function, are very considerably better than the numbers that you showed.

Nillius: I agree with you, of course, that the surgical results in small tumours are much better. But those tumours do not constitute a problem, as far as pregnancy is concerned. Also, it is not very good for the patients that there should be a general recommendation for surgery. I think that trans-sphenoidal microsurgery is a wonderful technique that surgeons should learn, but it should not be the first resort for young infertile women with prolactinomas, because it requires exceptional skill. The results from the departments in Paris, Munich, and Milan are really worrying: these are extremely interested and experienced departments, they have used the techniques for a long time, and still 40% of their micro-adenomas have hyperprolactinaemia postoperatively, plus another 7 patients a year later. We do not have any long-term follow-up on those patients. If every surgeon starts to operate, we could have terrible results. This factor must be pointed out.

Judd: I am not clear about patients who have been operated on, with a micro-adenoma, in whom the prolactin has been restored to normal, and everything appears to

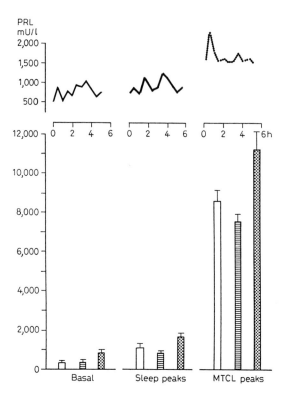

Fig. 1. Serum PRL levels in normal ovulatory women (open bars) and in infertile patients with normoprolactinaemic (hatched bars) and hyperprolactinaemic (dotted bars) corpus luteum insufficiency. Sleep peaks represent highest PRL levels observed during a night sleep period of at least 2 h (upper panel). PRL response to metoclopramide (MTCL) at +25 min are indicated. * denotes statistical significance from controls at $p < 0.05$ Reproduced with permission from *Bohnet and McNeilly* [1979].

be normal. What is the incidence of continuing infertility in that group? I have heard that there is an incidence which is higher than usual, but do you have any data?

Nillius: Pregnancy rates are not given in any of the surgical series. We are told how many achieve pregnancy, but not how many have attempted to become pregnant, so that you cannot get a rate. The only series which gives a pregnancy rate is the Paris series with a 42% pregnancy rate.

Friesen: I should like to ask Dr. *Nillius* if he has observed a patient who, during the course of pregnancy, spontaneously cured her prolactinaemia? Is it a common finding? Have you observed this phenomenon in many patients?

Nillius: We have had a few, but it is not a common finding. The vast majority have high prolactin levels after the pregnancy, and continue to have amenorrhoea.

Bohnet: We have, recently, been studying the response to metoclopramide during normal pregnancies, and during pregnancies complicated by tumours. So far, we have been able to study 4 patients with micro-adenomas during pregnancy, and 3 of them showed a normal response to metoclopramide, comparable to that seen in normal pregnancies. One patient complained of headaches and dizziness; X-ray of the sella did not reveal any major increment, but the prolactin response to metoclopramide was enhanced. After treatment with bromocriptine, the symptoms disappeared.

We have studied infertile patients, and we did metoclopramide tests. We found three patterns: (1) very normal, and this applies to the basal levels as well as to the responses when they are compared to the levels in normally cycling women; (2) elevated basal prolactin levels with normal response; (3) an exaggerated response in the follicular phase, as well as during the luteal phase. We were interested in whether these response patterns were of any physiological meaning. We compared the peak prolactin levels observed during sleep and correlated these with the peak levels observed after metoclopramide (fig. 1). The sleep peaks observed do correlate well with the metoclopramide-induced prolactin peaks (r is about 0.7, which is significant). If you have normal basal levels and response to metaclopramide, and you treat the patient with lisuride or bromocriptine, you have no amelioration of the luteal progesterone output. On the other hand, if you treat the other two classes of infertile patient with either an enhanced basal rate or an enhanced response to metoclopramide, luteal progesterone levels are improved. I think that the metoclopramide test is a good tool with which to monitor high prolactinaemia during pregnancy and also during corpus luteum insufficiency.

Reference

Bohnet, H.G. and McNeilly, A.S.: Prolactin: assessment of its role in the human female. Hormone metabol. Res. *11:* 533–546 (1979).

A New Approach to the Identification of Prolactin-Secreting Microadenomas?

Maurice F. Scanlon, Maria D. Rodriguez-Arnao, Alan M. McGregor, Mina Pourmand, David R. Weightman, David B. Cook, Antonio Gomez-Pan[1], Mark Lewis and Reginald Hall

Endocrine Unit, Departments of Medicine and of Clinical Biochemistry, Royal Victoria Infirmary, Newcastle-upon-Tyne

It is now widely recognised that hyperprolactinaemia in man may be due to prolactin (PRL)-secreting pituitary microadenomas [3, 12, 16, 18, 30, 33] which can cause diagnostic problems since the pituitary fossa may be entirely normal radiologically [13, 33] and the precise interpretation of minor radiological variations in the sella turcica is difficult. The inadequacy of current radiological diagnostic methods emphasises the importance of developing functional tests to determine the presence of a PRL-secreting microadenoma.

Since the dominant inhibitory hypothalamic control of PRL release is mediated by the central neurotransmitter dopamine (DA) [8, 21, 32] and thyroid-stimulating hormone (TSH) release is also under inhibitory hypothalamic dopaminergic control [27–29], it is of interest to ascertain whether any alteration in dopaminergic control of TSH release occurs in the presence of a primary disturbance in PRL secretion. We have therefore compared the pattern and degree of TSH and PRL responses to acute administration of the DA receptor-blocking drug domperidone in hyperprolactinaemic patients with and without radiological evidence of small prolactinomas and in normal subjects in order to define the dopaminergic control mechanism in these patients and to assess the potential value of any such alteration in dopaminergic tone in determining the presence of a PRL-secreting microadenoma. In view of the recent report [22] that one may be able to distinguish between adenomatous and non-adenomatous hyperprolactinaemia

[1] Departamento de Ensocrinologia de la Seguridad Social, Pabellon 8, Cuidad Universitaria, Madrid.

using the new drug nomifensine, an inhibitor of synaptic DA reuptake with mild DA agonist activity [26], we have administered this drug to a selected group of hyperprolactinaemic patients.

Patients and Methods

The following groups of subjects were studied: 9 normal euthyroid females (ages 19–48 years) with normal basal PRL levels (PRL<500 mU/l) and 31 hyperprolactinaemic patients (30 female, 1 male; ages 18–45 years). Of the hyperprolactinaemic patients, 12 had definite radiological abnormalities of the pituitary fossa consistent with the presence of a prolactinoma and the presence of a pituitary adenoma was confirmed by definitive neuro-radiological investigation with metrizamide cisternography. 3 patients (2 female, 1 male) with particularly large tumours had clinical and biochemical evidence of hypopituitarism and a further adenoma patient had persistent absence of the TSH response to TRH following successful treatment of hyperthyroidism due to Graves' disease 1 year previously. These 4 patients received nomifensine and placebo only. The remaining 19 hyperprolactinaemic patients had radiologically normal pituitary fossae.

Each subject whilst recumbent received domperidone (10 mg, intravenously; Janssen Pharmaceuticals, Beerse, Belgium) and on a separate occasion normal saline (0.9%; 2 ml, intravenously). 19 hyperprolactinaemic patients with abnormal fossae also received nomifensine (200 mg, orally, Hoechst Pharmaceuticals, UK, Ltd.) on a further occasion. Drug or placebo was administered at zero time following insertion of a forearm venous cannula 30 min previously. Blood was sampled at $-30, -15, 0, 10, 20, 30, 45, 60, 75, 90, 105$ and 120 min and serum was separated and stored at $-20°C$ until assayed. TSH was measured by a specific radio-immunoassay for human TSH [11] which has been optimised for sensitivity and precision [28] and using as standard MRC 68/38. Serum PRL was measured by standard radio-immunoassay [24] using WHO 75/504 as standard. TSH and PRL responses are expressed either as the incremental change at each time point or as the sum of increments above or below basal values from samples taken at each time point following drug or placebo administration. Results are expressed as the mean $\pm$ SE in mU/l for TSH and mU/l $\times 10^{-3}$ for PRL. Results were analysed statistically using both paired and non-paired 'Students tests'.

Results

TSH levels (sum of increments, mean $\pm$ SE, mU/l) rose significantly following domperidone in both normal females ($p < 0.001$) and in patients with biochemical and radiological evidence of prolactinomas ($p < 0.001$) although the rise was very much greater in prolactinoma patients than in normal females ($p < 0.001$, fig. 1, 3). However, the PRL response to DA

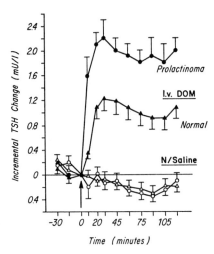

Fig. 1. Incremental TSH responses to domperidone (DOM, 10 mg, i.v., closed symbols) and N/Saline (0.9%, 2 ml, i.v., open symbols) in normal females and in patients with prolactinomas. Vertical bars represent mean ± SE.

antagonism was significantly less in prolactinoma patients than in normal females (p <0.01, fig. 3).

The hyperprolactinaemic patients with radiologically normal pituitary fossae can be clearly divided into two groups on the basis of the presence (n = 10) or absence (n = 9) of an exaggerated TSH response to DA antagonism (mean +2 SD of normal female response = 17.9 mU/l, figure 2). When grouped in this way radiologically normal hyperprolactinaemics with an exaggerated TSH response to DA antagonism show a significantly reduced PRL response when compared with both normal females (p < 0.01) and with radiologically normal hyperprolactinaemics who show a normal TSH response (p <0.01; fig. 3). As might be anticipated from this data there is a weak but significant inverse relationship between TSH and PRL responses to domperidone in the combined groups (n = 28, r = –0.39, p < 0.05). The degree of correlation improves when one compares the 10-min increment in TSH levels following domperidone with the total incremental PRL response to the same agent (n = 28, r = –0.47, p < 0.02).

Basal PRL levels were similar in the prolactinoma patients and radiologically normal hyperprolactinaemics with an exaggerated TSH response (5.8 ±1.2 vs 4.9 ±1.2; mean ± SE, mU/l+10-/) but were significantly

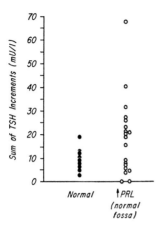

Fig. 2. Total incremental TSH response to domperidone in normal females and in hyperprolactinaemic females with normal pituitary fossae.

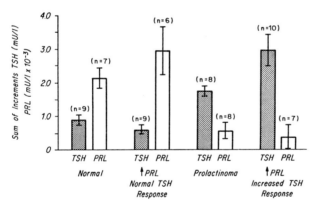

Fig. 3. Comparison of total incremental TSH and PRL responses to domperidone in normal and hyperprolactinaemic females. Vertical bars represent mean ± SE.

higher in the latter group than in radiologically normal hyperprolactinaemics with a normal TSH response to DA antagonism (4.9 ± 1.2 vs 1.5 ± 0.2; $p < 0.01$).

It should be emphasized that the variability and spontaneous fluctuation in basal PRL values was considerable in all groups particularly amongst those with unboubted prolactinomas. In this group 9 patients received nomifensine and 4 failed to show any degree of PRL decline over a 120 minute period. However, the decline in PRL levels in the 5 patients was slight and matched by an equal or greater decline in PRL levels after placebo. It certainly did not appear to result from any PRL suppressant activity of nomifensine. Overall, prolactinoma patients showed no significant changes after nomifensine or placebo (sum or PRL increments over 120', mean $\pm$ SE; mU/l $\times$ 10^{-3}: nomifensine 2.9 $\pm$ 2.8 vs. control $-3.4 + 2.7$).

Discussion

The finding of acute TSH release in all the subjects studied following administration of the novel DA receptor blocking drug domperidone, which does not cross the blood-brain barrier [2, 19, 25] confirms the inhibitory role of DA in the control of TSH release in man and indicates that DA is acting at the level of either the anterior pituitary or median eminence since these tissues lie outside the blood-brain barrier [4, 23].

Patients with prolactinomas show greater TSH release after domperidone than normal females whilst at the same time PRL release is reduced. Since the TSH and PRL are released as a result of blockade of the inhibitory dopaminergic mechanism, the degree of hormonal release reflects the degree of dopaminergic inhibition and therefore prolactinoma patients show increased dopaminergic inhibition of TSH release and reduced dopaminergic inhibition of PRL release when compared with normal females. This difference is seen as early as 10 min following domperidone administration. The reasons for such an alteration in dopaminergic control in man are unknown but there is persuasive animal evidence that PRL can regulate its own secretion by increasing hypothalamic dopaminergic activity [1, 5, 7, 9, 10, 14, 36]. The increased DA inhibition of TSH release in prolactinoma patients may therefore reflect the short loop, positive feedback action of the elevated circulating PRL levels on hypothalamic dopaminergic activity.

Since hyperprolactinaemia is sustained and the PRL response to DA antagonism may be reduced in patients with prolactinomas, an underlying defect in the specific dopaminergic inhibition of PRL release must be present, a finding in accord with the recent demonstration of a central dopaminergic defect in such patients [6, 35]. This is likely to be a local and specific defect since dopaminergic effects on normal pituitary tissue around the adenoma (as evidenced by the TSH response to DA antagonism) are increased. The nature and site of this local defect in the inhibitory dopaminergic control of PRL release is unknown. It is now well established that DA is the major physiological PRL-inhibiting factor acting directly on the lactotroph [21, 32] and the concentration of DA in hypophysial portal blood is sufficient to inhibit PRL release *in vivo* [8]. Since PRL levels can be effectively suppressed in prolactinoma patients by DA or DA agonist drugs [33], the defect may well be in the delivery of endogenous DA to the adenomatous lactotrophs possibly as a consequence of local intrapituitary disruption of portal microvascular relationships secondary to adenoma development and growth. It remains possible however that the adenomatous lactotroph is less sensitive to endogenous DA concentrations.

When hyperprolactinaemic patients with radiologically normal pituitary fossae are subdivided according to the presence or absence of an exaggerated TSH response to DA antagonism, some interesting parallels emerge. Patients with an exaggerated TSH response resemble prolactinoma patients in terms of both the basal PRL level and the reduction in the PRL response to DA antagonism whilst patients with normal TSH responses also show broadly normal PRL responses to DA antagonism.

By inference, patients with persistent hyperprolactinaemia and radiologically normal pituitary fossae who show an exaggerated TSH response to DA antagonism may well harbour PRL-secreting microadenomas, whereas patients with normal, reduced or absent responses may constitute the so-called 'functional' group in whom the aetiology of the hyperprolactinaemia remains unknown but in some of whom a diffuse reduction in inhibitory hypothalamic dopaminergic activity may be the primary aetiological event. Obviously in large tumours causing diffuse pituitary damage with loss of thyrotrophs this argument would not apply but this situation should not cause diagnostic problems since the lesion would be easily visible on skull X-ray.

Investigation of the altered dopaminergic control of normal pituitary tissue around prolactinomas could prove to be more reliable in excluding the presence of PRL-secreting microadenomas than assessment of the PRL response to dynamic testing with agents such as thyrotropin-releasing hormone (TRH) and DA antagonists such as metoclopramide and chlorpromazine which have proved less helpful than initially anticipated [13, 15, 20, 33, 34]. Whilst the absence of PRL responses to TRH and metoclopramide in a hyperprolactinaemic patient may well indicate an underlying prolactinoma [3, 16, 17], the presence of considerable PRL responses to these agents by no means excludes such a diagnosis [20, 33] in part because of the wide range of normal PRL responses to these agents. In contrast the normal range of TSH response to DA antagonism is relatively narrow.

We would suggest therefore that more attention be directed towards assessment of the altered dopaminergic control of normal pituitary function around prolactinomas and we would further suggest that the absence of an exaggerated response to DA antagonism in a patient with persistent hyperprolactinaemia and normal pituitary radiology is not compatible with the presence of any degree of lactotroph antonomy and may thus exclude the presence of a PRL-secreting microadenoma.

We found no significant suppression of PRL levels by nomifensine in the prolactinoma patients with large adenomas when considered as a group which supports the previous finding of others [22]. However, we experienced difficulty in determining the presence or absence of significant PRL suppression in some individuals at specific time points following nomifensine administration because of the spontaneous fluctuation in basal PRL levels. Further studies in combination with histology should reveal whether or not the same difficulties will be encountered with microadenomas.

Acknowledgements

We are grateful to the Scientific and Research Committee of the Newcastle Area Health Authority (Teaching) for financial support. Dr. *M.F. Scanlon* is currently a holder of a Harkness fellowship of the Commonwealth Fund of New York. Dr. *A.M. McGregor* is a holder of an MRC training fellowship in endocrinology.

References

1 Advis, J.P.; Hall, T.R.; Hodson, C.A.; Mueller, G.P., and Meites, J.: Temporal relationship and role of dopamine in 'short-loop' feedback of prolactin. Proc. Soc. exp. Biol. Med. *155:* 567–570 (1977).
2 Costall, B.; Fortune, D.H., and Naylor, R.J.: Neuropharmacological studies on the neuroleptic potential of domperidone (R33812). J. Pharm. Pharmac. (in press, 1979).
3 Cowden, E.A.; Ratcliffe, J.G.; Thomson, J.A.; MacPherson, P.; Doyle, D., and Teasdale, G.M.: Tests of prolactin secretion in diagnosis of prolactinomas. Lancet *i:* 1155–1158 (1979).
4 Dobbing, J.: The blood-brain barrier. Physiol. Rev. *41:* 130–188 (1961).
5 Eikenburg, D.C.; Ravitz, A.J.; Gudelsky, G.A., and Moore, K.E.: Effects of estrogen on prolactin and tubero-infundibular dopaminergic neurons. J. neural Transm. *40:* 235–244 (1977).
6 Fine, S.A. and Frohman, L.A.: Loss of central nervous system component of dopaminergic inhibition of prolactin secretion in patients with prolactin secreting pituitary tumors. J. clin. Invest. *61:* 973–980 (1978).

7   Fuxe, K.; Hökfelt, T., and Nilsson, O.: Factors involved in the control of the activity of the tubero-infundibular dopamine neurons during pregnancy and lactation. Neuroendocrinology 5: 257–270 (1969).
8   Gibbs, D.M. and Neill, J.D.: Dopamine levels in hypophysial stalk blood in the rat are sufficient to inhibit prolactin secretion *in vivo*. Endocrinology 102: 1895–1900 (1978).
9   Grandison, L.; Hodson, C.; Chen, H.T.; Advis, J.; Simpkins, J., and Meites, J.: Inhibition by prolactin of post-castration rise in LH. Neuroendocrinology 23: 312–317 (1977).
10 Gudelsky, G.A.; Simpkins, J.; Mueller, G.P.; Meites, J., and Moore, K.E.: Selective actions of prolactin on catecholamine turnover in the hypothalamus and on serum LH and FSH. Neuroendocrinology 22: 206–215 (1976).
11 Hall, R.; Amos, J., and Ormston, B.J.: Radioimmunoassay of human serum thyrotrophin. Br. med. J. i: 582–585 (1971).
12 Hardy, J.; Beauregard, H., and Robert, F.: Prolactin-secreting pituitary adenomas: trans phenoidal microsurgical treatment; in Robyn and Harter, Progress in prolactin physiology and pathology, pp. 361–369 (Elsevier/North-Holland, Amsterdam 1978).
13 Healy, D.L.; Pepperell, R.J.; Stockdale, J.; Bremner, W.J., and Burger, H.G.: Pituitary autonomy in hyperprolactinaemic secondary amenorrhoea: results of hypothalamic-pituitary testing. J. clin. Endocr. Metab. 44: 809–819 (1977).
14 Hökfelt, T. and Fuxe, K.: Effect of prolactin and ergot alkaloids in the tuberoinfundibular dopamine (DA) neurons. Neuroendocrinology 4: 100–122 (1972).
15 Jacobs, H.S.; Franks, S.; Murray, M.A.F.; Hull, M.G.R.; Steele, S.J., and Nabarro, J.D.N.: Clinical and endocrine features of hyperprolactinaemic amenorrhea. Clin. Endocrinol. 5: 439–454 (1976).
16 Jaquet, P.; Grisoli, F.; Guibout, M.; Lissitzky, J.-C., and Carayon, P.: Prolactin secreting tumours: endocrine status before and after surgery in 33 women. J. clin. Endocr. Metab. 46: 459–466 (1978).
17 Kleinberg, D.L.; Noel, G.L., and Frantz, A.G.: Chlorpramazine stimulation and L-dopa suppression of plasma prolactin in man. J. clin. Endocr. Metab. 33: 873–876 (1971).
18 Kleinberg, D.L.; Noel, G.L., and Frantz, A.G.: Galactorrhoea: a study of 235 cases, including 48 with pituitary tumors. New Engl. J. Med. 296: 589–600 (1977).
19 Laduron, P. and Leysen, J.: Domperidone, a novel gastrokinetic and anti-nauseant drug, interacts selectively with dopamine receptors. 7th Int. Congr. of Pharmacology, Paris 1978, p. 34, No. 71.
20 Lamberts, S.W.J.; Birkenhäger, J.C., and Kwa, H.G.: Basal and TRH-stimulated prolactin in patients with pituitary tumours. Clin. Endocrinol. 5: 709–711 (1976).
21 Macleod, R.M. and Lehmeyer, J.E.: Studies on the mechanism of the dopamine-mediated inhibition of prolactin secretion. Endocrinology 94: 1077–1085 (1974).
22 Muller, E.E.; Genazzani, A.R., and Murru, S.: Nomifensine: diagnostic test in hyperprolactinaemic states. J. clin. Endocr. Metab. 47: 1352–1357 (1978).
23 Oldendorf, W.H.: Brain uptake of radiolabelled amino acids, amines and hexoses after arterial injection. Am. J. Physiol. 221: 1629–1639 (1971).
24 Reuter, A.M.; Kennes, F.; Gevaert, Y., and Franchimont, P.: Homologous radioimmunoassay for human proclatin. Int. J. nucl. Med. Biol. 3: 21–28 (1976).
25 Reyntgens, A.J.; Niemegeers, C.J.E.; Van Neuten, J.M.; Laduron, P.; Heykants, J.;

Schellekens, K.H.L.; Marsbloom, R.; Jageneau, A.; Broekaert, A., and Janssen, P.A.J.: Domperidone, a novel and safe gastrokinetic anti-nauseant for the treatment of dyspepsia and vomiting. Arzneimittel-Forsch. *28:* 1194–1196 (1978).

26 Scanlon, M.F.; Gomez-Pan, A.; Mora, B.; Cook, D.B.; Dewar, J.H.; Hildyard, A.; Weightman, D.R.; Evered, D.C., and Hall, R.: Effects of nomifensine, an inhibitor of endogenous catecholamine reuptake, in acromegaly, in hyperprolactinaemia, and against stimulated prolactin release in man. Br. J. clin. Pharmacol. *4:* 1915–1975 (1977).

27 Scanlon, M.F.; Mora, B.; Shale, D.J.; Weightman, D.R.; Heath, M.; Snow, M.H., and Hall, R.: Evidence for dopaminergic control of thyrotrophin (TSH) secretion in man. Lancet *ii:* 421–423 (1977).

28 Scanlon, M.F.; Weightman, D.R.; Shale, D.J.; Mora, B.; Heath, M.; Snow, M.H., and Hall, R.: Dopamine is a physiological regulator of thyrotrophin (TSH) secretion in man. Clin. Endocrinol. *10:* 7–15 (1979).

29 Scanlon, M.F.; Rodriguez-Arnao, M.D.; Pourmand, M.; Weightman, D.R.; Shale, D.J.; Lewis, M., and Hall, R.: Catecholaminergic interactions in the regulation of TSH secretion in man. J. Endocr. Invest. (in press, 1980).

30 Sherman, B.M.; Harris, C.E.; Schlechte, J.; Dvello, T.M.; Halmi, N.S.; Vangilder, J.; Chapler, F.K., and Granner, D.K.: Pathogenesis of prolactin-secreting pituitary adenomas. Lancet *ii:* 1019–1021 (1978).

31 Swanson, H.A. and DuBoulay, G.: Borderline variants of the normal pituitary fossa. Br. J. Radiol. *48:* 366–369 (1975).

32 Takahara, J.; Arimura, A., and Schally, A.V.: Suppression of prolactin release by a purified porcine PIF preparation and catecholamines infused into a rat hypophyseal portal vessel. Endocrinology *95:* 462–465 (1974).

33 Thorner, M.O.: Prolactin: clinical physiology and the significance and management of hyperprolactinaemia; in Martini and Besser, Clinical neuroendocrinology, pp. 319–361 (Academic Press, London 1977).

34 Tolis, G.; Somma, M.; Van Campenhout, J., and Friesen, H.: Prolactin secretion in sixty-five patients with galactorrhoea. Am. J. Obstet. Gynec. *118:* 91–101 (1974).

35 Van Loon, G.R.: A defect in catecholamine neurones in patients with prolactin-secreting pituitary adenoma. Lancet *ii:* 868–871 (1978).

36 Voogt, J.L. and Carr, L.A.: Plasma prolactin levels and hypothalamic catecholamine synthesis during suckling. Neuroendocrinology *16:* 108–118 (1974).

M.F. Scanlon, MD, Department of Medicine, Welsh National School of Medicine, Heath Park, Cardiff, (Wales)

Discussion

Adler: It would be important to point out that your changes in TSH require that you have a very good TSH assay, and that most clinically available TSH assays are not going to detect 1 milliunit/litre changes in the basal levels and may or may not be able to detect the kinds of changes that you saw.

Scanlon: One certainly does need a sensitive TSH assay. However, the sort of responses that we were seeing in the radiologically normal hyperprolactinaemics with an exaggerated TSH response were big responses in a significant number. These would be detectable on routine TSH assays, although many would be excluded. The other point is that there may well be ways of exaggerating this difference. We know, from previous data of our own, that if one gives a shot of this agent at 11 p.m., as opposed to 11 a.m., the TSH response is three or four times greater. Another possible way of enhancing the difference would be to give TRH, maximally or at a submaximal dose, 30–60 min after the dopamine antagonist drug. We have previously shown that there is slight enhancement in normal subjects of the TSH response to TRH following acute administration of dopamine antagonist drugs.

Eastman: In view of these findings that it was very difficult to correlate prolactinomas or tumours with abnormal X-rays, and also with prolactin levels, do you think that what you may be looking at may be some functional effect at that time? I wonder, therefore, whether this is reproducible. Have you looked at these patients over a longer period of time? In other words, do they move from one group to another?

Scanlon: No, we have not looked over a longer period of time. All the prolactinoma group had abnormal fossa on plain skull X-ray, and in several of them that was confirmed with further definitive neuroradiological studies.

Harding: You showed the abnormal TSH responses and this was more associated with the higher basal prolactin levels. How clear-cut was that distinction? Has it enabled you to assign a weight to a particular level of prolactin – intermediate between normal and grossly elevated?

Scanlon: At the moment, the division is artificial, but we are impressed by the parallel in terms of prolactin responses and basal prolactin levels, when we do divide the groups in that way.

Cowden: We also, in patients who have persistent hyperprolactinaemia but quite normal pituitary radiology, have been able to identify two groups. One of these had quite abolished prolactin responses to TRH and metoclopramide, and an absent diurnal rhythm. The other group was clinically indistinguishable, but they had an obvious, although blunted, prolactin response to TRH and metoclopramide.

Hershman: I am troubled by the concept that you have established increased dopaminergic inhibition of TSH release from this test with a single dose. Perhaps you could tell us something about the dose response of normals to domperidone. Is there an alteration in the dose-response curve in the patients with prolactinomas?

Scanlon: There is no alteration in dose-response curve in patients with prolactinomas. In fact, the 10-mg dose that we use was established on the basis of a dose-response study done on normal males and females, which shows clearly that a 10-mg dose produces maximal effect. Increasing the dose to 15 or 20 mg produces similar effects to the 10-mg dose. There is no evidence from our studies at those dosages that domperidone is suddenly developing a partial agonist action at the higher dosage.

Metrizamide Cisternography in the Assessment of Pituitary Tumour Size, with Special Reference to the Demonstration of Tumour (Prolactinoma) Shrinkage on Bromocriptine Therapy

Keith Hall, Alan M. McGregor, Maurice F. Scanlon and Reginald Hall[1]

Department of Neuroradiology, Regional Neurological Centre, Newcastle General Hospital, and Endocrine Unit, Department of Medicine, Royal Victoria Infirmary, Newcastle-upon-Tyne

Metrizamide (Amipaque), a non-ionic water-soluble contrast medium, which was developed by Nyegaard & Co. of Oslo for radiculography and myelography, is now being used in some centres for the visualisation of the intracranial subarachnoid cisterns. This is termed metrizamide cisternography (when plane radiographs of the cisterns are obtained) or metrizamide CT cisternography (when computerised tomography, or CT, is used). These metrizamide studies have proved of great value, by visualising the suprasellar cistern, in the investigation of lesions in and near the pituitary fossa [1, 3, 4, 7, 11]. Indeed, because we find these studies to be simpler to perform and also better tolerated than pneumoencephalography in the investigation of such lesions, the latter procedure has been abandoned for over 2 years. Plane skull radiography and CT scanning, with cerebral angiography in some cases, are still essential preliminary investigations in all such cases. One undoubted value of the metrizamide study is that, because it is so well tolerated by patients, repeat investigations can be readily performed [3, 9].

The clinical and biochemical effects of different types of treatment used for pituitary tumours are quite readily assessed. On the other hand, the radiological demonstration of a change in size of such a lesion after this treatment may be more difficult. A decrease in the size of the sella on repeat skull radiographs has been used by some [12], but differing degrees of magnification on the films can give spurious results. Repeat CT scanning has proved of value in a few cases [6, 8] but, as a rule, the only tumours which

---

[1] The authors wish to thank Miss *A. Kellett* for photographic assistance.

are definitely shown to change in size by this technique are the large ones with prominent extrasellar extensions. A small, purely intrasellar tumour is difficult to demonstrate on the CT scan and a change in size is almost impossible to define. Repeat pneumoencephalography [2] is rarely performed, being so unpleasant for the patient in many cases, and for this reason we have been investigating the use of repeat metrizamide studies, especially following the treatment of pituitary tumours with the drug bromocriptine.

Some of the results in this paper, as well as details of many of the technical aspects of the metrizamide studies which we perform, have been described elsewhere [3, 4]. This short paper will concentrate mainly on the patients with prolactinomas and the results of the repeat studies, following bromocriptine.

Patients and Methods

Over 150 metrizamide studies have been performed, so far, in our centre for lesions in or near the pituitary fossa, mostly for known or suspected pituitary tumours. A detailed analysis of the findings in the first 100 such examinations, in 85 patients, has been made elsewhere [3]. In this group of patients were 35 prolactinomas, 19 acromegalics, 10 chromophobe adenomas and 2 patients with Nelson's syndrome.

In all cases a plane skull radiograph (lateral and PA) was obtained first, one of the criteria for inclusion of the patient in this first 100 metrizamide examinations being that the sella should be definitely abnormal, with enlargement or asymmetry, on these films. Tomography of the sella turcica was not routinely carried out. A conventional CT scan, without and with intravenous contrast medium administration, was performed before the metrizamide study in all cases. In fact, the only absolute contraindication to doing a metrizamide study is the presence of a large suprasellar extension of a pituitary tumour. If this completely fills the suprasellar cistern then the metrizamide cannot cover it and the procedure is of no value. In addition, if there is a very large suprasellar mass that is obstructing the foramen of Monro and causing hydrocephalus, the lumbar puncture to introduce the metrizamide is potentially dangerous.

The principle of the metrizamide study is to outline the suprasellar cistern with the contrast medium [3, 4]. This is done by running it up into the head, from the lumbar region where it is introduced, by tilting the prone patient head-down by about 60°. The cistern, its contents and extensions are imaged by lateral radiography (the metrizamide cisternogram) and by CT scanning (the metrizamide CT cisternogram). Usually, the CT scans are obtained in both the axial (horizontal) and coronal planes and from all these images a full three-dimensional appreciation of the pituitary tumour and its extensions can be built up. Adjacent structures such as the optic chiasm are well-shown by this technique and it is very easy to fill an empty sella (intrasellar arachnoid herniation) with the metrizamide dropping into it in the prone position that the patient adopts.

All the patients had had a full clinical and biochemical work-up, including examination of the visual fields. The 35 patients proven to have prolactinomas all had typical clinical features, with amenorrhoea, galactorrhoea and sometimes infertility in the 22 females, and impotence, and occasionally galactorrhoea, in the 13 males. A few had headache and/or visual disturbances as well. In all, there was a moderate or marked elevation of the serum prolactin level, as assessed by radio-immunoassay, the venous blood samples being taken 30 min after cannulation. 11 of these prolactinoma cases had had previous therapy, namely surgery and often radiotherapy, some years before. Such treatment had also been used in 5 acromegalics and 3 with chromophobe adenomas.

A large number of all the patients with pituitary tumours were subsequently treated with the drug bromocriptine for 3 months to 1 year, the dose being 20 mg daily. After the treatment, repeat radiological investigations (skull radiograph, CT scan and metrizamide study) were performed in 12 patients, with 2 undergoing two repeats. 7 of these 12 were patients with prolactinomas, the remaining 5 being 3 acromegalics (2 having two repeats), 1 chromophobe adenoma and 1 Nelson's syndrome. The 7 prolactinoma patients (4 male and 3 female) had an age range of 23–60 years (mean 40 years) and all had marked hyperprolactinaemia, the range being 2,100–9,150 mU/l (mean 7,450). The females had amenorrhoea and galactorrhoea and the males impotence, with galactorrhoea in one and headache in another. 2 of these 7 had had previous surgery and postoperative radiotherapy, for presumed chromophobe adenomas, 7 and 3 years before.

Results

As aready stated, most of the results of these metrizamide studies are given elsewhere [3]. Only the facts concerning the prolactinomas and the repeat studies will be detailed here, with further amplification.

Prolactinomas

The findings are summarised in table I. All the patients had an abnormal sella on plane skull radiography. The CT scan could demonstrate a large suprasellar extension, and even a moderate sized one, quite easily, but a small one was often difficult to show. Similarly, a large empty sella could often be well-shown but smaller ones were more difficult to definitely identify, especially with the artefacts that occur in the sellar region. The metrizamide study was found to give a much better delineation of a suprasellar extension or an empty sella, as well as the normal structures in and near the suprasellar cistern, than these other radiological techniques.

As will be noted from table I, there was a strikingly high incidence of an empty sella in patients with prolactinomas, being seen in 16 cases (46%). Previous treatment, by surgery and radiotherapy, could only be implicated as a cause for this in 5 patients. Thus, most were primary empty sellae, the prolactinoma usually being a microadenoma despite the definitely abnormal sella which is often taken to indicate that there is probably a large tumour

Table I. Findings on metrizamide studies in 35 prolactinomas (figures in parenthese are numbers of patients who had had previous surgery and radiotherapy)

Suprasellar extension		7 (1)	(20%)
Small	3		
Moderate	3 (1)		
Large	1		
Normal		12 (5)	(34%)
Empty sella		16 (5)	(46%)
Small	3		
Moderate	6 (2)		
Large	7 (3)		

Table II. Changes detected on repeat metrizamide studies in 7 patients with prolactinomas following bromocriptine therapy (Rx) (SSE = suprasellar extension; ES = empty sella)

Patient	Before Rx	After Rx
1	moderate SSE	small ES
2	moderate SSE	small ES
3	small SSE	small ES
4	small SSE	small ES
5	normal	small ES
6	normal	small ES
7	small ES	moderate ES

present, a macroadenoma (diameter greater than 10 mm). Incidentally, an empty sella was also seen in 6 of the 19 patients with acromegaly (32%), though only 3 of these were of the primary type. Only 2 patients with chromophobe adenomas had an empty sella (20%) and both of these were the secondary type, the result of previous treatment.

Results of Repeat Radiological Studies, Especially Metrizamide Studies

All 12 of the patients who underwent repeat radiological investigation after treatment with bromocriptine showed definite clinical improvement, as well as a reduction in the appropriate increased hormone levels. This was especially marked in the 7 prolactinoma cases, the prolactin levels falling to a mean value of 530 mU/l.

There was no detectable change in the appearance of the sella on the repeat skull radiographs, and only in one prolactinoma, with a suprasellar extension, was there a suggestion of a change in size on the repeat CT scan,

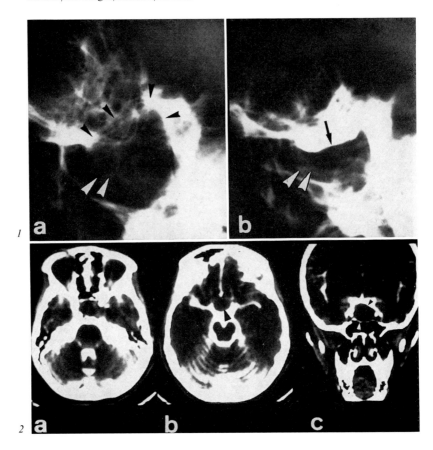

Fig. 1. Metrizamide cisternogram in patient 1, showing: (a) moderate suprasellar extension of prolactinoma (black arrowheads) and (b) small empty sella (black arrow), after 4 months of bromocriptine therapy (the asymmetrical sella floor is shown by white arrowheads).

Fig. 2. Metrizamide CT cisternogram in patient 1, before bromocriptine therapy: (a, b) axial plane scans showing moderate suprasellar extension of prolactinoma (black arrowhead in b); (c) coronal plane scan showing extension (small black arrowhead) and sella floor (small white arrowhead).

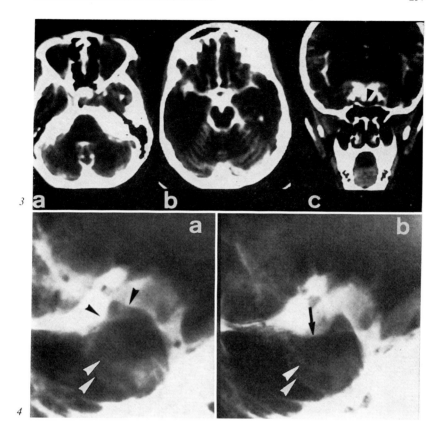

Fig. 3. Metrizamide CT cisternogram in patient 1, after 4 months of bromocriptine therapy; (a, b) axial plane scans showing normal suprasellar cistern on *b* and probable metrizamide in sella on *a*; (c) coronal plane scan confirming partial empty sella (black arrowhead), and disappearance of suprasellar extension.

Fig. 4. Metrizamide cisternogram in patient 3, showing: (a) small suprasellar extension of prolactinoma (black arrowheads); (b) small empty sella (black arrow), after 3 months of bromocriptine therapy (the asymmetrical sella floor is shown by white arrowheads).

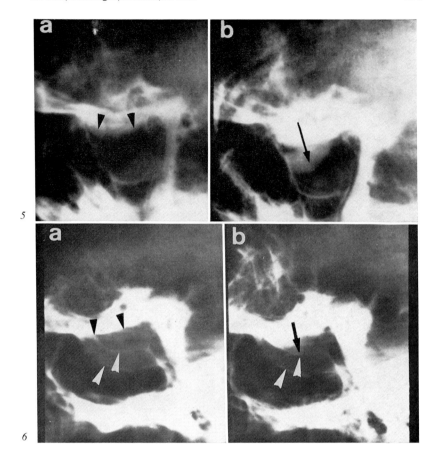

Fig. 5. Metrizamide cisternogram in patient 7, showing: (a) small empty sella (black arrowheads); (b) moderate empty sella (black arrow), after 12 months bromocriptine therapy.

Fig. 6. Metrizamide cisternogram in patient 5, showing: (a) normal position of diaphragma sellae (black arrowheads); (b) small empty sella (black arrow), after 12 months of bromocriptine therapy (the asymmetrical sella floor is shown by white arrowheads).

and this was by no means definite. The metrizamide study was the only method used which could definitely demonstrate the subtle changes in tumour size which were discovered. In fact, only the prolactinomas exhibited any change and this was seen in all 7 patients; the other 5 tumours did not alter at all, despite the clinical and biochemical improvements. The changes seen in the prolactinomas are summarised in table II and some examples shown in figures 1–6.

Discussion

In the planning of any type of treatment of a pituitary tumour many surgeons and radiologists feel that the information derived from a skull radiograph, even with tomography, and a conventional CT scan is inadequate. It is usually essential to show precisely the relationship of any tumour to the adjacent structures, especially the optic chiasm and carotid arteries, and the surgeon will usually like to know that there is an empty sella if he plans a trans-sphenoidal approach.

Cerebral angiography will obviously demonstrate the adjacent arteries and, in many centres, the pneumoencephalogram is still used to define the suprasellar extension of a pituitary tumour and show its relationship to the optic chiasm. Metrizamide studies do not appear to be widely used, as yet, to investigate pituitary lesions, but we feel that this type of investigation has many advantages over pneumoencephalography and now no lnger perform the latter.

One of the advantages of the metrizamide study is that it is simpler to perform and probably quicker as well. We have found that the side-effects are fewer than with the pneumoencephalogram and those that do occur come on after the whole procedure is complete. This is in contradistinction to the pneumoencephalogram during which headache, nausea and vomiting, sometimes with hypotensive episodes as well, quite frequently cause the procedure to be abandoned. This has never happened with any of our metrizamide studies. The combination of the metrizamide cisternogram and the metrizamide CT cisternogram (in both the axial and coronal planes) builds up a three-dimensional image of the whole pituitary and juxta-pituitary region. Finally, being well-tolerated, the metrizamide study can be repeated to assess the results of treatment, as in this small study. The only real disadvantage of metrizamide is its high cost, but perhaps this can be reduced as it is more widely used.

An empty sella is a common finding, as has been stressed [5], but this fact still seems to be poorly recognised by many clinicians. It is seen in normal persons and we have shown an incidence of 20% in a small group of 25 patients who had a metrizamide cisternogram following on from a cervical myelogram [4]. An empty sella can also be seen in cases of pituitary tumour, the most widely recognised type being the secondary one which follows previous treatment, especially a combination of surgery and postoperative radiotherapy. 19 of the 66 pituitary tumour cases in this series had had previous treatment of this type and 10 (53%) were shown to have an empty sella, usually moderate or large in size.

The combination of a primary empty sella with a pituitary tumour is less-well recognised and the reports of sporadic cases seem to emphasise how uncommon it is. We have found it to occur in 31% of prolactinomas and 9% of acromegalics, and quite often it is large in size. This is a point that will be of great importance to the surgeon planning a trans-sphenoidal removal of any pituitary tumour so that a postoperative leak of CSF can be avoided. It is possible that the metrizamide study will demonstrate an empty sella more readily than the pneumoencephalogram, for the metrizamide tends to drop into the intrasellar arachnoid herniation very easily in the prone position. On the other hand, a special inverted 'hanging-head' position, usually with sagittal plane tomography, has to be adopted during the pneumoencephalogram to displace the CSF in the empty sella by the lighter air.

It could be argued that the pneumoencephalogram would have shown the fairly subtle changes in tumour size, sometimes being detected only by changes in the position of the diaphragma sellae, which the repeat metrizamide studies demonstrated in all seven prolactinomas treated with bromocriptine. There was no definite change in tumour size detectable by the other radiological methods used. The only way to prove conclusively which was the best would be to do both, but this is not ethically justifiable. We feel that the metrizamide study is the best one available at the moment to show these subtle changes in tumour size. Possibly the third and fourth generation CT scanners will be able to show intrasellar detail well-enough to detect these changes, though the radiation dose to the patient with multiple overlapping thin-section slices will be high.

Bromocriptine therapy has a pronounced effect, both clinically and biochemically, on patients with different types of pituitary tumours with restoration of normal function in many instances [10]. The only tumour which have been shown to definitely decrease in size radiologically have been prolactinomas [9], as we have shown here. This is despite some clinical evidence of decreasing tumour size in acromegaly, chromophobe adenomas and Nelson's syndrome in which there has been improvement in the visual fields [9]. The reason for this is not clear but may be related to the dose of bromocriptine used and the time interval before follow-up radiological investigation. In the two instances in which we used repeat metrizamide studies on two occasions, in patients with acromegaly, we assumed that the time interval would have been adequate but still no change in size was detectable.

References

1   Drayer, B.P.; Rosenbaum, A.E.; Kennerdell, J.S.; Robinson, A.G.; Bank, W.O., and Deeb, Z.L.: Computed tomographic diagnosis of suprasellar masses by intrathecal enhancement. Radiology *123:* 339–344 (1977).
2   Ezrin, C.; Kovacs, K., and Horvath, E.: Hyperprolactinaemia; morphologic and clinical considerations. Med. Clins N. Am. *62:* 393–408 (1978).
3   Hall, K.: Metrizamide studies of the sellar region. Clin. Radiol. (in press).
4   Hall, K. and McAllister, V.L.: Metrizamide cisternography in pituitary and juxtapituitary lesions. Radiology *134:* 101–108 (1980).
5   Kaufman, B. and Chamberlain, W.B.: The ubiquitous 'empty' sella turcica. Acta radiol. (Diag.) *13:* 413–425 (1972).
6   Landolt, A.M.; Wüthrich, R., and Fellman, H.: Regression of pituitary prolactinoma after treatment with bromocriptine. Lancet *i:* 1082–1083 (1979).
7   Manelfe, C.; Guiraud, B.; Espagno, J., and Rascol, A.: Cisternographie computerisée au métrizamide. Revue neurol. *134:* 471–484 (1978).
8   McGregor, A.M.; Scanlon, M.F.; Hall, K.; Cook, D.B., and Hall, R.: Reduction in size of a pituitary tumour by bromocriptine therapy. New Engl. J. Med. *300:* 291–293 (1979).
9   McGregor, A.M.; Scanlon, M.F.; Hall, R.; Cook, D.B., and Hall, K.: Effects of bromocriptine on pituitary tumour size. Br. med. J. *ii:* 700–703 (1979).
10 Parkes, D.: Drug therapy: bromocriptine. New Engl. J. Med. *301:* 873–878 (1979).
11 Sheldon, P. and Molyneux, A.: Metrizamide cisternography and computed tomography for the investigation of pituitary lesions. Neuroradiology *17:* 83–87 (1979).
12 Wass, J.A.H.; Moult, P.J.A.; Thorner, M.O.; Dacie, J.E.; Charlesworth, M.; Jones, A.E., and Besser, G.M.: Reduction of pituitary tumour size in patients with prolactinomas and acromegaly treated with bromocriptine with or without radiotherapy. Lancet *ii:* 66–69 (1979).

Keith Hall, MB, Department of Neuroradiology, Regional Neurological Centre, Newcastle General Hospital, Westgate Road, Newcastle-upon-Tyne NE4 6BE (England)

Discussion

Jacobs: I have heard of some rather disquieting side-effects, or complications, and I wondered whether you would comment on what the side-effects of metrizamide cisternography have been in your experience.

McGregor: Obviously, the major effect to which you are alluding is seizure activity and epilepsy. We have now performed studies on 200 patients. In no patient has there ever been any suggestion of an epileptic seizure. 3 of the patients had had epilepsy following surgical treatment, and they were covered through metrizamide by Valium® and their usual medication without any problem. The complications that are associated with the investigation are headaches; there is no doubt that the patients do get headaches and do feel nauseated afterwards, but it is nothing like the effects of encephalography. We now have patients who have had as many as four metrizamide cisternograms, with no ill-effect whatsoever. We premedicate our patients with metoclopramide, and we give them an oral or intramuscular analgesic afterwards, if it is indicated.

Jacobs: I take it you have had no blindness?

McGregor: I have never heard of that as a complication, we certainly have not seen it.

Jacobs: We have seen 1 case of sudden blindness, which was transient thankfully.

Thorner: I believe that the use of metrizamide with CT, certainly with the low resolution scanner that we have, does not really contribute very much. The tomography with metrizamide is far more helpful.

McGregor: We have not, actually, as yet used metrizamide with tomography. In the 16 patients whom we have followed through bromocriptine, we have found that we get very adequate delineation of the tumour just with CT scanning.

Buckman: The incompetent diaphragma sella is an extremely common finding in autopsy studies. Your incidence rate of 60% for empty sella, do you attribute that to the incidence among the normal population, or do you think that it has something to do specifically with the presence of prolactinomas?

McGregor: I am certainly aware of the post-mortem data. I would suspect that we are just looking at a variant of the normal which has nothing at all to do with the pathology in the pituitary fossa. It is very much easier to see it with metrizamide than it is with air.

Frantz: Have you done any of these studies following pituitary surgery? Do you find the same depression and tendency to empty sellas that you do after bromocriptine?

McGregor: I should have said that, in 7 of the 49 patients with large tumours that we studied, studies done immediately after surgery in 2 of them certainly delineated fairly closely an area of empty fossa, as well as showing an area in which tumour was still present.

del Pozo: We have performed studies on rat-induced pituitary tumours. The concomitant treatment with bromocriptine led to a reduction of tumour size, a considerable reduction of weight, and then an actual reduction of lactotrope cells, as identified by immunohistochemistry. Electron microscopy studies gave no evidence of cellular damage, which means that the reduction of the tumour is by actual inhibition of reproduction of this particular cell-line, since there was no evidence of necrosis, or any abnormality of intercellular function.

Jacobs: I think that this might be an appropriate time to raise the question of the reasons for the discrepancies that we have heard in findings about the efficacy of various therapies. *McGregor* has demonstrated very beautifully the reduction in size of pituitary adenomas with bromocriptine. We heard from *Nillius*'s comprehensive and elegant review of the less than desirable results of trans-sphenoidal surgery, even with microadenomas, and even in the hands of experienced and enthusiastic surgeons. Obviously, one possible mechanism for the, let us say, poor results of surgery may be that considerable tissue, hypersecreting prolactin, resides not in the micro-adenoma, that is often removed successfully, but in the surrounding pituitary tissue. The concept of hyperplasia of the lactotrope cell, not a new idea, is one that most of us probably had in mind in thinking about those results. Clearly, bromocriptine, unless it wipes out lactotrope cell populations completely, is subject to the same kind of hazard, if that is what is going on. Long-term follow-up of all therapeutic groups is required. I should like to throw out the notion of hyperplasia in the surrounding remaining tissue for general discussion, and only add the general comment that in the autopsy series in which a 25% prevalence of prolactin cell hyperplasia was quite readily identified, about one third of the glands harbouring hyperplactic lactotropes also harboured micro-adenomas.

Thorner: I fully agree with *Jacobs* about the concept that we might not be dealing with a solitary micro-adenoma in the pituitary. But I want to go back to the point of the long-term follow-up of various different forms of treatment. Although it is, obviously, necessary to do studies of long-term follow-up after bromocriptine treatment, the experience, we must not forget, is now built up over nearly 10 years, and to my knowledge no patient who has responded to bromocriptine initially has failed to respond to the long-term treatment. Also, one has to say that, certainly in our experience, withdrawal of bromocriptine in almost all cases leads to a recurrence of hyperprolactinaemia. I can only remember one patient who, following withdrawal of bromocriptine, had normal prolactin levels which remained down for a long period.

Buckman: The concept of hyperplasia is a very real one, even in the absence of discrete adenomas. Last year, in the course of some 6 weeks, we subjected 3 patients to trans-sphenoidal pituitary exploration. 2 of the patients had radiographic evidence of macro-adenomas, the third had a micro-adenoma. We removed tissue that appeared to be grossly abnormal, and yet after careful sectioning, the neuropathologist has assured us that no tumour was present in 2 of the individuals: the final histological diagnosis was atypical hyperplasia. In the third, no abnormality whatsoever was found. We extracted portions of each of these tissues, and found greatly elevated prolactin concentrations in each, suggesting that, indeed, the lactotropes are hyperfunctioning in those instances. Following surgery, serum prolactin levels decreased in each of the individuals, but failed to return to normal. Perhaps it is a continuum in the sense that there is a phase of hyperactivity and normal histology, then areas of hyperplasia, and finally the evolution of a pituitary adenoma.

del Pozo: In Zurich, *Landolt* usually removes the adenoma, then takes a little biopsy of the surrounding tissue. We have a recent case of a young girl; the adenoma was removed, and the biopsy of the normal tissue showed a number of active prolactin cells. What was unexpected was that the prolactin in the periphery was 3 ng/ml, and was not reacting to TRH – which is a common finding following surgery. After 1 month, basal prolactin was 4 ng/ml and normally responding to TRH. One should be cautious in interpreting any findings from biopsy during surgery, because the histological appearance might not agree with the real secretion rates of the tissue.

Primary Hypothyroidism: Abnormalities of Prolactin Secretion and Pituitary Function

Loren I. Dolman

Letterman Army Medical Center, Presidio of San Francisco, Calif. and University of Utah Medical Center, Salt Lake City, Utah

The coexistence of TSH hypersecretion and pituitary enlargement frequently occurs in patients with primary hypothyroidism [1, 7, 10]. Measurement of prolactin (PRL) prompted by the presence of galactorrhea in some of these patients has shown elevations [7, 9], but in larger series basal PRL is inconsistently elevated [1, 6, 11].

Some patients show PRL hyperresponsiveness to TRH [1, 7] (which may revert somewhat after *L*-thyroxine therapy [7]), and when examined, PRL suppression after *L*-dopa has been reported as normal [7].

Other endocrine function including GH and cortisol response to insulin-induced hypoglycemia may also be abnormal [1], sometimes reverting post-therapy.

We have examined 8 patients referred to us for pituitary evaluation because of 'abnormal sella turcica' on X-ray. They later turned out to have primary hypothyroidism, and we report the evaluation of PRL dynamics and pituitary function studies in this group.

Materials and Methods

The 8 patients included 2 males and 6 females, 5 of the latter complaining of galactorrhea. All were previously felt by the referring physician to have abnormal sellar X-rays, some felt to have hypothyroidism, and all worked up according to an extensive pituitary protocol. All medications were discontinued 2 weeks prior to admission.

Baseline studies drawn fasting at 8 a.m. the first morning included T_4, TSH, antithyroid antibodies, PRL, GH, LH, FSH, and testosterone. The patients were given 3 g of metyrapone orally at midnight, and cortisol, 11-deoxycortisol, and ACTH obtained the next 8 a.m. as previously described [4]. TRH, 500 µg i.v., was given and serial samples

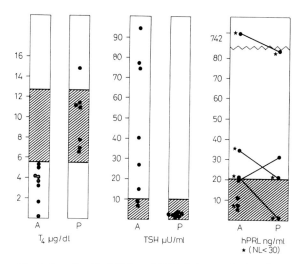

Fig. 1. Baseline T_4, TSH, and PRL before and after therapy. Shaded area indicates normal range. A = before and P = after *L*-thyroxine replacement.

drawn for TSH and PRL. *L*-dopa, 500 mg p.o., was given and serial samples obtained for GH and PRL.

Visual field examination was obtained by the Ophthalmology Service in most cases. All patients had sellar tomography and review of sellar films by a neuroradiologist. After diagnosis of primary hypothyroidism and completion of pituitary workup, patients were replaced on *L*-thyroxine (100–200 μg orally per day) and pituitary studies repeated 6 weeks to 6 months post-normalization of T_4 and TSH. All hormones were measured by radioimmunoassay.

Results

Most of the patients were thought to have borderline to moderate hypothyroidism on physical examination. Visual fields obtained in 7 patients were normal. Upon review skull X-rays were thought to be abnormal in 4 patients, with another 3 showing abnormalities in sellar tomography. The eighth patient was considered to have normal sellar tomography.

Baseline T_4 ranged from 0.4 to 4.8 μg/dl with TSH 6.3 to 94 μU/ml (fig. 1) considered diagnostic of primary hypothyroidism in all cases. Antithyroid antibodies were present in all patients measured. After therapy, T_4 and TSH were normal (except in 1 patient where T_4 was slightly elevated due to excessive *L*-thyroxine replacement, fig. 1).

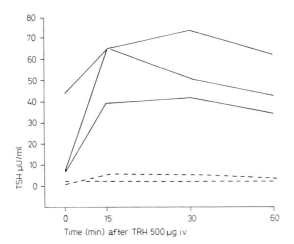

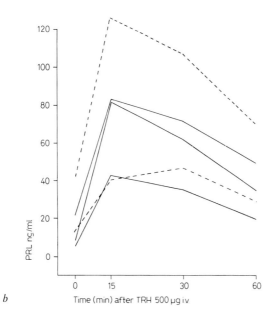

Fig. 2a, b. TSH and PRL following administration of TRH. —— = Before and --- = after L-thyroxine replacement.

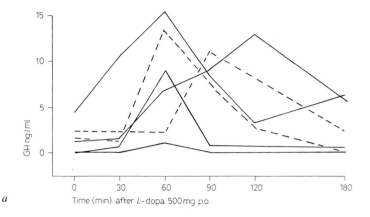

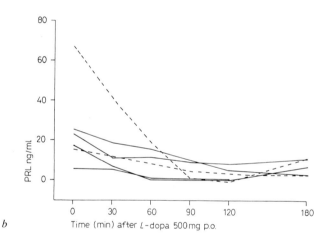

Fig. 3a, b. GH and PRL following administration of *L*-dopa.

Baseline PRL in 4 patients (normal < 30 ng/ml) was 6.1, 20.8, 34.4, and 742.0, respectively, and < 2, 21.3, and 84.5 in the latter 3 patients post-therapy. In 3 other patients, measured by a different assay (normal PRL < 20 ng/ml), PRL was 5.9–19.8 prior to therapy. In the 1 patient initially having PRL of 19.8, a post-therapy PRL value of 34 ng/ml was obtained (fig. 1).

After TRH, TSH showed a supernormal response in 3 patients tested (fig. 2), 2 of them showing a flattened response post-therapy. PRL (fig. 2) showed a supernormal rise in 3 patients after TRH. After L-thyroxine therapy, the PRL rise somewhat normalized in 1 patient, but was even more exaggerated in another (the patient mentioned previously with mildly elevated baseline PRL post-therapy) (fig. 2).

After L-dopa, 3 of 4 patients had normal GH elevation (fig. 3). The one abnormal response was obtained in a patient with consistently normal PRL measurements. PRL after L-dopa (fig. 3) dropped normally in all patients tested before and after therapy.

ACTH and 11-deoxycortisol after metyrapone were abnormal in 2 of the 8 patients prior to therapy, normalizing in 1 of them after thyroid replacement. In the former patient, testosterone was 239 ng/dl prior to therapy (normal for males >300), normalizing to 956 ng/dl post-replacement. The other patients were considered to have normal gonadotrophic function. In 1 female (patient with the highest PRL), LH and FSH were 2.5–4.9 mIU/ml (normal < 30) both before and after therapy; however, FSH stimulated to 8.8 mIU/ml after clomiphene citrate (50 mg orally twice a day for 2 weeks), considered a low-normal response.

Discussion

Pituitary enlargement secondary to primary hypothyroidism is not an uncommon phenomenon. Indeed in animal models [3] and in humans [10] this is due to hyperplasia of TSH-secreting cells in the pituitary in response to decreased feedback from thyroid hormones. In studies of sellar volume, sellar enlargement was frequently found in patients with milder forms of primary hypothyroidism and was inversely correlated with thyroid hormone levels [1].

Visual fields are usually normal in patients with primary hypothyroidism as was the case in all our patients throughout their course. An exception is when the pituitary enlargement is sufficient to disrupt the sellar architecture and disturb the optic chiasm [8]. One report describes deterioration of visual fields after L-thyroxine replacement [12].

Plain skull films may be mildly abnormal in up to 31.7% of normal patients [13] and sellar tomography in 17–42%, depending on the series [2, 13]. In our patients these abnormalities were the basis for referral. Perhaps accurate sellar volume measurements may be more specific in assessing sellar enlargement.

In all patients, primary hypothyroidism was diagnosed by a combination of inappropriately high TSH with low T_4. All patients had complete suppression of TSH after L-thyroxine replacement, and normalization of T_4 except in the 1 patient mentioned above on excessive replacement.

Baseline PRL was not usually elevated in our patients in agreement with the published series, but did drop to normal from an elevated level post-therapy. In the patient with very high PRL prior to therapy, PRL dropped after 6 months of replacement but was still four times normal, suggesting the presence of a prolactin-secreting pituitary adenoma. In the patient with high-normal PRL prior to therapy and abnormal PRL post-therapy (on repeated determinations), a small autonomous PRL-secreting pituitary adenoma is suspected. This is also suggested by her supernormal PRL response to TRH post-therapy. Whether autonomous prolactin-secreting adenomata are more common in patients with primary hypothyroidism or their presence in our 2 patients simply reflects the 9% incidence of prolactin-containing adenomata reported recently in a autopsy series of normals [2] cannot be determined by our small study.

Normal suppression of PRL after L-dopa in some of our patients with primary hypothyroidism is in agreement with the previously mentioned report [7] and is probably more common than in series of patients with prolactin-secreting pituitary adenomata where good PRL suppression after L-dopa is often but not always the rule [5]. GH response to L-dopa paralleled that found in a normal population.

Although reduction of GH and cortisol responses to hypoglycemia has been reported in primary hypothyroidism [1], ACTH determinations have not been consistently obtained to rule out the possibility of primary adrenal failure in association with Hashimoto's thyroiditis. In our patients, 2 did have significant ACTH deficiency prior to therapy, 1 normalizing after adequate L-thyroxine replacement. In addition 1 patient had gonadotropin deficiency prior to therapy, and another was borderline. Both these findings suggest some degree of pituitary insufficiency in these patients, due to increased sellar size from hyperplastic thyrotrophs. (Note: the patient with low testosterone had a normal PRL, implying that hyperprolactinemia was not the cause of his gonadal failure.) Total pituitary failure has been appreciated in patients with primary hypothyroidism [14], but its frequency is difficult to assess as complete pituitary function testing is not available in all the patients reported.

Conclusions

We agree that pituitary abnormalities are a common finding in primary hypothyroidism if one accepts abnormal sellar X-rays for diagnosis. Baseline PRL and suppression after *L*-dopa are most often normal. The frequency of galactorrhea and supernormal response of PRL to TRH probably implies underlying hypersecretion of PRL, however. Some patients may actually have a PRL-secreting pituitary adenoma concomitant with primary hypothyroidism, not resolving after adequate *L*-thyroxine replacement. We suggest PRL measurements in all patients with primary hypothyroidism after adequate *L*-thyroxine replacement. The dynamic tests of PRL release appear to be of no diagnostic value. Finally, pituitary insufficiency (usually partial) is evident in some of these patients and probably should be diagnosed early. We suggest testing for adrenal and gonadal reserve (and GH reserve in children) after an adequate period of *L*-thyroxine replacement.

References

1 Bigos, S.T.; Ridgway, E.C.; Kourides, I.A., and Maloof, F.: Spectrum of pituitary alterations with mild and severe thyroid impairment. J. clin. Endocr. Metab. *46:* 317–325 (1978).
2 Burrow, G.N.; Wortzman, G.; Rewcastle, N.B., and Holgate, R.C.: Hyperprolactinemia, abnormal sellar tomograms and pituitary adenomas in an unselected autopsy series. Program and Abstracts of the 6th Int. Congr. of Endocrinology, Melbourne 1980, p. 402.
3 Dent, J.N.; Gadsden, E.L., and Furth, J.: Further studies on induction and growth of thyrotropic pituitary tumors in mice. Cancer Res. *16:* 171–174 (1956).
4 Dolman, L.I.; Nolan, G., and Jubiz, W.: Metyrapone test with adrenocorticotrophic levels. Separating primary from secondary adrenal insufficiency. J. Am. med. Ass. *241:* 1251–1253 (1979).
5 Dolman, L.I.; Roberts, T.S., and Tyler, F.H.: Secretory pituitary adenomata: diagnosis and follow-up after selective transsphenoidal pituitary adenectomy. Present Concepts intern. Med. (US Army) *13:* 33–82 (1980).
6 Honbo, K.S.; Van Herle, A.J., and Kellett, K.A.: Serum prolactin levels in untreated primary hypothyroidism. Am. J. Med. *64:* 782–787 (1978).
7 Keye, W.R.; Yuen, B.H.; Knopf, R.F., and Jaffe, R.B.: Amenorrhea, hyperprolactinemia, and pituitary enlargement secondary to primary hypothyroidism. Successful treatment with thyroid replacement. Obstet. Gynec., N.Y. *48:* 697–702 (1976).
8 Leiba, S.; Landau, B., and Ber, A.: Target gland insufficiency and pituitary tumors. Acta endocr., Copenh. *60:* 112–120 (1960).
9 Pelosi, M.A.; Langer, A.; Zanvettor, J., and Devanesan, M.: Galactorrhea associated with primary hypothyroidism. Report of two cases. Obstet. Gynec., N.Y. *49:* 12s–14s (1977).

10 Samaan, N.A.; Osborne, B.M.; Mackay, B.; Leavens, M.E.; Duello, T.M., and Halmi, N.S.: Endocrine and morphologic studies of pituitary adenomas secondary to primary hypothyroidism. J. clin. Endocr. Metab. *45:* 903–911 (1977).
11 Sheeler, L.R.; Kumar, M.S., and Schumacher, O.P.: Serum prolactin levels in hypothyroidism. Cleveland Clin. Q. *45:* 307–310 (1978).
12 Stockigt, J.R.; Essex, W.B.; West, R.H.; Murray, R.M.L., and Breidahl, H.D.: Visual failure during replacement therapy in primary hypothyroidism with pituitary enlargement. J. clin. Endocr. Metab. *43:* 1094–1100 (1976).
13 Swanson, H.A. and Boulay, G. du: Borderline variants of the normal pituitary fossa. Br. J. Radiol. *48:* 366–369 (1975).
14 Tunbridge, W.M.G.; Marshall, J.C., and Burke, C.W.: Primary hypothyroidism presenting as pituitary failure. Br. med. J. *i:* 153–154 (1973).

L.I. Dolman, MD, Department of Medicine, The Permanente Medical Group, 2200 O'Farrell, San Francisco, CA 94115 (USA)

Discussion

Adler: It has been my experience that the galactorrhoea seen sometimes in hypothyroidism, like so many other clinical abnormalities with hypothyroidism, takes a long time to resolve. It would be very interesting to restudy your patients 1 or 2 years after full replacement therapy. You may find that most of those abnormalities have returned to normal.

Frantz: Do you have any idea what the true incidence of sellar enlargement is in primary hypothyroidism? Because that is not at all an uncommon condition, especially in an elderly population, and we usually just do not do sellar X-rays in these cases. My impression is that it is not terribly common. There have been several articles which have commented on it, but I think that it must be a small minority of patients with hypothyroidism that exhibit frank sellar enlargement.

Dolman: Obviously, in the hundreds of patients whom we see with primary hypothyroidism, we do not do it either; these 8 slipped through the protocol.

Adler: In the paper by *Bigos et al.*, they clearly showed a very good correlation between the severity of the hypothyroidism, and the size by Dicureau's formula of the sella. I think that there is a good correlation between sellar volume and severity of hypothyroidism, but where normal becomes abnormal might be in question.

Hershman: There was a paper by *Yamada et al.* in the *Journal of Clinical Endocrinology and Metabolism (42:* 817, 1976) showing a tight correlation between the log of the serum TSH concentration and the volume of the sella. So one might expect to see bigger sella enlargement with more severe hypothyroidism.

Jacobs: Although basal prolactin levels are most often in the normal range in adults with hypothyroidism, and the prevalence of abnormal sellar size, as *Frantz* pointed out, is not known with certainty, but may be small, both of these are somewhat altered in juvenile hypothyroidism, where the prevalence of basal hyperprolactinaemia is higher, and it is my impression, on an anecdotal basis, that the prevalence of abnormal sellar profiles on lateral skull films is also higher.

Prolactin and Cyclicity in Polycystic Ovary Syndrome

Role of Estrogens and the Dopaminergic System

Emilio del Pozo and Paolo Falaschi

Universitäts-Frauenklinik, Basle, and Clinica Medica V, University of Rome

The etiology of abnormal gonadotropin and sex steroids secretion found in patients with polycystic ovary syndrome (PCO) is not clearly understood. Although testing of LH feedback and hypothalamic control mechanisms discloses a sensitive pituitary gland in this condition, the primary disturbance that allows episodic LH secretion in the absence of cyclic activity remains unexplained. Furthermore, the role of peripheral organs in the maintenance of the vicious circle leading to excessive production of sex steroids needs to be assessed.

Recent knowledge of the role of prolactin (PRL) in human reproduction and of the control of gonadotropin release by dopaminergic mechanisms has amplified our understanding of this problem. In the present paper the possible role of PRL in the physiopathology of PCO is studied. The implication of dopamine receptors in the anomalous LH production in this syndrome will also be reviewed.

PRL and PCO: Response to TRH Stimulation and Dopaminergic Blockade

Enhanced estrogen formation from excessive ovarian or adrenal androgen production is a hallmark of PCO. Previous work by various investigators has demonstrated a higher incidence of elevated PRL in PCO [4, 5] probably due to the stimulatory effect of estrogens on pituitary lactotropes. Plasma measurements in 47 cases of PCO, selected on the basis of abnormally elevated LH and raised plasma levels of estrone and androgenic precursors, revealed an incidence of hyperprolactinemia of 27%. This elevation,

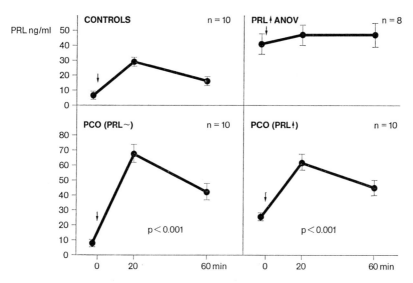

Fig. 1. PRL response to 200 μg TRH in PCO patients as compared with women with hyperprolactinemic anovulation (ANOV) and normal controls (means ± SE). From reference 6, by permission.

however, was only moderate, most values being between 15 and 30 ng/ml (N < 15 ng/ml), possibly reflecting a mild stimulatory effect of sex steroids. Similar plasma PRL levels are recorded in the early postconceptional time in which lactotropic stimulation by placental estrogens is just starting.

Based on pregnancy studies and on the analysis of situations of estrogen dominance, it was shown that opposite to pathologic low-estrogen hyperprolactinemia, TRH stimulation or dopaminergic blockade disclosed a clear positive PRL response in states of hyperestrogenism. Since PCO is characterized by a relative estrogen excess, the PRL response to TRH and to the dopaminergic antagonist haloperidol was compared with the type of reaction recorded in normal female volunteers and in hyperprolactinemic galactorrhea-amenorrhea [6]. As expected, the response curves to these agents were blunted in the idiopathic hyperprolactinemic patients (fig. 1). The PCO patients, however, hyperreacted to TRH irrespective of the basal PRL concentrations. The response was significantly higher ($p < 0.001$) than in control subjects. Haloperidol elicited an excessive PRL discharge only in hyperprolactinemic PCO patients ($p < 0.001$) (fig. 2). It was concluded that the priming effect of estrogens in PCO has a bearing on the type of response elicited.

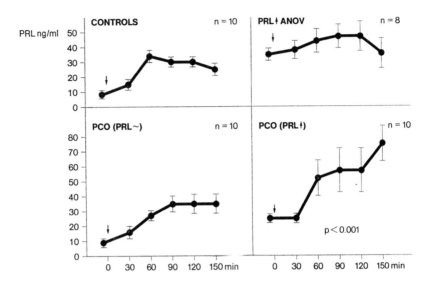

Fig. 2. Same as figure 1 using 1 mg haloperidol as stimulating agent (means ± SE). From reference 6, by permission.

In another experiment, *Falaschi et al.* [7] evaluated the pituitary PRL response to different stimuli involving the dopaminergic system. Three groups of 10 subjects each were selected: women with pituitary prolactinomas, hyperprolactinemic PCO patients with normal roentgenograms of the sella turcica, and normal female volunteers. The PRL response to nomifensine (DA reuptake-inhibitor), and domperidone (peripheral DA-receptor blocker) was assessed. In PCO patients PRL levels were significantly reduced by nomifensine, as observed in normal volunteers, but their response to domperidone was excessive when compared with the control subjects. Patients with prolactinomas failed to respond to the stimuli. These results lend further support to the hypothesis that in PCO syndrome hyperprolactinemia seems to be the functional consequence of estrogenic priming.

Effect of Bromocriptine Treatment on the Anovulation of PCO

Thorner et al. [17] initially reported that the reduction of PRL secretion in galactorrhea patients with sclerocystic ovaries resulted in restoration of ovulatory cycles. In a preliminary study, *Rocco et al.* [15] studied the effect of bromocriptine (Parlodel®), a potent dopamine agonist, in hyperprolac-

Table I. Basal data and results of bromocriptine treatment in hyperprolactinemic PCO (n = 10; mean ± SE). Adapted from ref. 15.

	PRL ng/ml	T ng/100 ml	BBT	Ovulation	
				n	P, ng/ml
Basal	29.7 ± 3.1	107.8 ± 7.0	M (10)	0	2.2 ± 0.4
Bromocriptine	4.2 ± 0.4	64.2 ± 5.1	B (8)	8	15.2 ± 1.8

PRL = Prolactin; T = testosterone; BBT = basal body temperature; M = monophasic; B = biphasic; P = progesterone (mean of 3 midluteal values).

Table II. Integrated LH plasma profiles in 4 subjects with PCO before and after bromocriptine treatment (Total mIU/ml/time)

Basal	Bromocriptine
2,632	960
2,205	667
3,738	787
3,420	990
	$p < 0.05$

tinemic PCO. The criteria used for selection were: (a) enlarged ovaries on laparotomy, laparoscopy or pelvigraphy; (b) high pulsatile LH without ovulatory surge and normal or low FSH; (c) raised levels of testosterone and estrone, with normal or low estradiol. Out of 10 patients studied, 3 had amenorrhea and 7 varying degrees of oligomenorrhea. Basal body temperature records were monophasic in all. Bromocriptine was administered in daily doses of 3.75–7.5 mg for several months. Cumulative data on plasma PRL, testosterone and estrogens are depicted in table I. Biphasic basal temperatures were recorded in 8 cases and ovulation was supported by plasma progesterone estimates in all of them (table I). Results of integration of LH profiles in 3 cases of this series and from another patient following the same treatment schedule are presented in table II. Bromocriptine not only lowered basal LH values but also significantly reduced the amplitude of the pulsatile waves.

The mechanisms involved in the origin of hyperprolactinemia of PCO are far from clear. Two hypotheses can be proposed based on the PRL

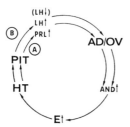

Fig. 3. Possible sites of action of bromocriptine in PCO: *A* Via PRL and adrenal androgens. *B* Via dopaminergic LH reduction. AD/OV = adrenals/ovary; AND = androgens; E = estrogens; HT = hypothalamus; PIT = pituitary.

action on adrenal androgen synthesis as discussed in the previous section and on the possible alteration of central dopaminergic mechanisms. Thus, the events leading to the classic clinical and biochemical picture of PCO are depicted in figure 3: exaggerated hypothalamic stimulation leads to inappropriately elevated circulating LH and subsequently to increased ovarian androgen synthesis in addition to an adrenal component. Target tissues transform androgens, mainly androstenedione into estrone, which in turn may disturb hypothalamic cyclicity. However, it cannot be ruled out that androgenic steroids may interfere with estrogen receptor synthesis to 'masculinize' the hypothalamus and produce a male type of LH-secreting pattern with wide pulsatile episodes and without cycling activity. On the other hand, pituitary exposure to augmented circulating estrogens may prime lactotrope cells to oversecrete PRL. This lactogen would in turn enhance dehydroepiandrosterone synthesis by the adrenals, an effect clearly demonstrated in hyperprolactinemia [2, 3]. Therefore, this mechanism would be rendered inoperative under PRL suppression with bromocriptine. However, its validity can be criticized in view of the difficulty in discriminating between abnormal androgen synthesis from inappropriate adrenal stimulation by PRL or due to the basic biochemical and enzymatic abnormalities inherent in PCO. In addition, excessive pituitary estrogenization is only relative in this particular situation: estrone is a metabolite of low estrogenic activity and, accordingly, basal PRL levels could not be correlated with estrone concentrations in PCO patients [*Falaschi and del Pozo,* unpublished]. *Kandeel et al.* [11] found no PRL stimulation by this steroid when administered to normally cycling women, whereas estradiol showed an immediate effect. A relative estradiol dominance, however, through reduced binding to circulating globulin remains a possibility. On the other hand, the effect of androgens on PRL secretion has not been clearly established although acute testosterone treatment has failed to modify PRL levels in normal volunteers despite a significant fall in plasma LH [13], in agreement with recent data published by *D'Agata et al.* [1]. However, prolonged exposure of the pituitary gland to exogenous testosterone has reportedly stimulated PRL secretion in rats [8]. Thus, enhancement of PRL release after chronic stimulation by androgens in PCO remains another possibility.

The dopaminergic theory offers an attractive alternative for the normalization of hypothalamic cyclic activity after bromocriptine administration. Recent studies have suggested that DA and DA agonists exert an inhibitory effect on gonadotropin and PRL release in humans [12] and that this effect can be correlated with endogenous estrogen levels [9]. Although chronic dopaminergic stimulation with bromocriptine has failed to alter the cyclic pattern of LH [14, 16], the basic situation may be different in PCO. Indeed, recent work by *Yen* [18] has revealed a prompt decline in LH levels in response to a 4-hour DA infusion. Considering the enhanced LH sensitivity to LHRH characteristic of PCO, a working hypothesis was advanced in that a concomitant increase in LHRH neuronal activity brings about a decrease in the inhibitory influence of hypothalamic DA [10]. Thus, the modulating effect of bromocriptine on LH in PCO could involve a dopaminergic mechanism unrelated to the well-known inhibition of PRL secretion. The fact that 42% of normoprolactinemic PCO patients have responded to bromocriptine with a fall in plasma testosterone [15] further supports this hypothesis.

References

1 D'Agata, R.; Gulizia, S.; Vicari, E.; Aliffi, A., and Polosa, P.: Effect of androgens on prolactin (PRL) release in humans. Acta endocr., Copenh. *90:* 409–413 (1979).
2 Bassi, F.; Giusti, G.; Vorsi, L.; Cattaneo, S.; Giannotti, P.; Forti, G.; Pazzagli, M.; Vigiani, C., and Serio, M.: Plasma androgens in women with hyperprolactinemic amenorrhea. Clin. Endocrinol. *6:* 5–10 (1977).
3 Carter, J.N.; Tyson, J.E.; Warne, G.L.; McNeilly, A.S.; Faiman, C., and Friesen, H.G.: Adreno-cortical function in hyperprolactinemic women. J. clin. Endocr. Metab. *45:* 973–980 (1977).
4 Duignan, N.M.: Polycystic ovarian disease. Br. J. Obstet. Gynaec. *83:* 593–602 (1976).
5 Falaschi, P.; Frajese, G.; Rocco, A.; Toscano, V., and Sciarra, D.: Polycystic ovary syndrome and hyperprolactinemia. J. Steroid Biochem. *8:* xiii (1977).
6 Falaschi, P.; Pozo, E. del; Rocco, A.; Toscano, V.; Petrangeli, E.; Pompei, P., and

Frajese, G.: Control of prolactin release in the polycystic ovary syndrome. Obstet. Gynec., N.Y. *55:* 579–582 (1980).

7   Falaschi, P.; Rocco, A.; Pompei, P.; Pozo, E. del, and Frajese, G.: High incidence of hyperprolactinemia in polycystic ovary (PCO) syndrome: functional or tumoral origin. 6th Int. Congr. of Endocrinology, 1980, p. 641, abstr.

8   Heras, F. de las and Negro-Vilar, A.: Effect of aromatizable androgens and estradiol on prolactin secretion in prepuberal male rats. Arch. Androl. *2:* 135–139 (1979).

9   Judd, S.J.; Rakoff, J.S., and Yen, S.S.C.: Inhibition of gonadotropin and prolactin release by dopamine: effect of endogenous estradiol levels. J. clin. Endocr. Metab. *47:* 494–498 (1978).

10 Judd, S.J.; Rigg, L.A., and Yen, S.S.C.: The effects of ovariectomy and estrogen treatment on the dopamine inhibition of gonadotropin and prolactin release. J. clin. Endocr. Metab. *49:* 182–184 (1979).

11 Kandeel, F.; Butt, W.R., and London, D.R.: Regulation of serum prolactin levels by oestrone and oestrone sulphate in normally menstruating women. Acta endocr., Copenh. suppl. *212:* 42 (1977).

12 Lachelin, G.C.L.; Leblanc, H., and Yen, S.S.C.: The inhibitory effect of dopamine agonists on LH release in women. J. clin. Endocr. Metab. *44:* 728–732 (1977).

13 Martin-Perez, J.; Pozo, E. del; Clarenbach, P.; Brownell, J., and Derrer, F.: Lack of dependency between androgen production and prolactin in man. Acta endocr., Copenh. suppl. 225, p. 174 (1979).

14 Pozo, E. del; Goldstein, M.; Friesen, H.; Brun del Ré, R., and Eppenberger, U.: Lack of action of prolactin suppression on the regulation of the human menstrual cycle. Am. J. Obstet. Gynec. *123:* 719–723 (1975).

15 Rocco, A.; Falaschi, P.; Pompei, P.; Pozo, E. del, and Frajese, G.: Chronic anovulation in polycystic ovary syndrome: role of hyperprolactinemia and its suppression with bromocriptine: IV International Symposium on pediatric and adolescent gynaecology (in press).

16 Schulz, K.-D.; Geiger, W.; Pozo, E. del, and Kunzig, H.J.: Pattern of sexual steroids, prolactin, and gonadotropic hormones during prolactin inhibition in normally cycling women. Am. J. Obstet. Gynec. *132:* 561–566 (1978).

17 Thorner, M.O.; McNeilly, A.S.; Hagen, C., and Besser, G.M.: Long-term treatment of galactorrhea and hypogonadism with bromocriptine. Br. med. J. *ii:* 419–422 (1974).

18 Yen, S.S.C.: The polycystic ovary syndrome. Clin. Endocrinol. *12:* 177–207 (1980).

E. del Pozo, MD, Sandoz AG, CH-4002 Basle (Switzerland)

Discussion

Adler: Is there something that distinguishes those patients with mild hyperprolactinaemia from those who do not have it? Did you try bromocriptine in those patients who were normal prolactinaemics with PCO?

del Pozo: The cases studied up to now are not sufficient to find a correlation. We could think that the cause for the hyperprolactinaemia was the elevated oestrogens, especially oestrone; we are studying more cases, perhaps we shall get a better correlation. It is significant so far. Patients with normal prolactinaemia were also treated. There was a 40% response in these people, only observed in the patients responding with a fall in androgens. For that reason, we think that 'demasculinization' of the hypothalamus may have induced cycling again. This is not specially related to prolactin; prolactin was normal in this group.

Beumont: In contrast to *del Pozo*'s work on the hyperoestrogen state, anorexia nervosa is a condition found in young females in which there is self-induced weight loss, accompanied by low LH levels and low oestrogen levels. We studied the response to LH-RH in a group of emaciated anorexia nervosa patients. Some patients appeared to have a prolactin response to LH-RH during refeeding: 9 patients who were actually cooperating in our programme and rapidly gaining weight, at the level of 1 kg per week or more, appeared to show a prolactin response to LH-RH stimulation, while the patients who were not gaining weight were no different to our control subjects.

Ajayi: We have about 14 patients with hyperprolactinaemia who were put on bromocriptine. We did blood values of DHEA-S and testosterone and progesterone. Then, we did some stimulation tests on them with dexamethasone and ACTH. If you follow them for 3–4 weeks, you will find that the Δ^4 metabolites are increased, and at the same time you find a decrease in the Δ^5 metabolites. It may be that the isomerization activity is being inhibited by prolactin. Is this a possible effect of prolactin?

del Pozo: I do not know. We were trying to disclose an effect of bromocriptine at the level of the adrenals, for instance. *Neher and Podesta* were unsuccessful at finding an effect of this drug on an *in vitro* system of dispersed adrenal cells. I am sorry that I cannot comment on the other point, we did not do it.

Jacobs: This talk of a short luteal phase and prolactin concentrations reminds me that when *Shearman and Korenman* reported, some time ago, on reproductive senescence, they found short luteal phases quite common in the perimenopausal period. I wondered if anyone has systematic data, just for nosological purposes, on prolactin levels in perimenopausal women?

del Pozo: At ages 40–50 and 60 in women, the prolactin values are decreasing parallel to the decrease in ovarian function as estimated by basal plasma estradiol.

Prolactin Response as a Guide to Neuroleptic Treatment of Schizophrenia

Larry E.J. Evans, Peter J.V. Beumont, B. Luttrell, A.J. Henniker and J. Penny

Department of Psychiatry, University of Queensland University of Sydney, and The Department of Nuclear Medicine, Royal North Shore Hospital, Sydney

Galactorrhoea has been recognized as a side-effect of neuroleptic medication since the introduction of these drugs into clinical practice in the 1950s [1]. At the time, the pathogenesis of the phenomenon was not udnerstood, and its possible relevance to psychiatry was not recognized. It is now realized that antipsychotic drugs cause galactorrhoea because they block the dopamine-mediated inhibition of prolactin release [2].

All antipsychotic drugs, with the possible exception of clozapine, cause a rise in plasma prolactin, and there appears to be a high correlation between a drug's antipsychotic potency and its ability to disinhibit prolactin release. On the other hand, most psychotropic drugs which do not have an antipsychotic effect do not cause a significant rise in prolactin. One of the few exceptions is sulpiride (Eglonyl®), a benzamide antidepressant, which is related pharmacologically to metoclopramide (Maxillon®).

The most widely accepted biological theory concerning the causation of schizophrenia is that the illness is related to overactivity of dopaminergic neurones in the mesolimbic area of the brain. According to this hypothesis, the antipsychotic effect of neuroleptic drugs is dependent on their ability to block dopamine in this area [3]. The high correlation between an individual drug's antipsychotic and prolactin elevating effects suggests that the dopamine receptors involved in these 2 actions may be similar [4]. For these reasons it has been suggested that the prolactin response may be a useful model of antipsychotic effect, and that prolactin values may be used as an *in vivo* index of the amount of neuroleptic actually reaching the brain [5]. If this were the case, it would be of great practical usefulness, as most antipsychotic drugs have many biologically active metabolites.

Haloperidol, a butyropherone neuroleptic and a potent dopamine receptor blocker, is an effective antipsychotic agent. It is unusual among neuroleptics in that it does not appear to have biologically active metabolites [6]. The purpose of our present study has been to examine the relationship between plasma haloperidol levels, plasma prolactin and therapeutic response in a group of disturbed schizophrenics treated with haloperidol.

Material and Methods

We studied a group of 17 patients, 9 males and 8 females, with a mean age of 43 years. All patients initially underwent a 2-week placebo drug washout period. They were then treated with haloperidol 10 mg/day for 8 weeks. Psychiatric response was monitored clinically, and by use of the In-patient Multi-dimensional Psychiatric Scale [7], an accepted measure of psychiatric response. At the end of 8 weeks, 7 patients were judged to have made a satisfactory clinical response and were not studied further. The remaining 10 patients, 5 males and 5 females (mean age 44 years), were treated for a further period of 10 weeks with a dose of haloperidol which increased by 10 mg/day every week up to a maximum dose of 80 mg/day. Psychiatric status was assessed throughout, and all patients showed a marked improvement by the end of the trial.

Blood samples were drawn weekly, at 9.00 a.m., 12 h after the last dose of haloperidol, and assayed for plasma haloperidol and plasma prolactin. Haloperidol was measured by a gas chromatograph electron capture technique [8]. Prolactin was measured by radio-immunoassay [9] using MRC reference standard A 71/222. The interassay coefficient of variation of the prolactin assay was 6.9%.

Results

In the patients who responded, plasma prolactin rose from a level of 383 μIU/ml during washout to 1,017 μIU/ml at the end of 8 weeks, a highly significant change ($p < 0.02$). In the non-responders, plasma prolactin levels did not change significantly. It is worth noting that average plasma haloperidol levels at the end of the 8 weeks was 4.9 ng/ml in the responders, but only 2.8 ng/ml in the non-responders. There was, however, considerable overlap between the groups in haloperidol levels. While plasma prolactin values were a predictor of good clinical response, plasma haloperidol was not, at least at the low range of dosage.

In figure 1 mean plasma prolactin and plasma haloperidol values are shown throughout the period of study. A significant rise of plasma prolactin and plasma haloperidol during the fixed, low-dosage period of study was found only in those patients who made a satisfactory clinical response to

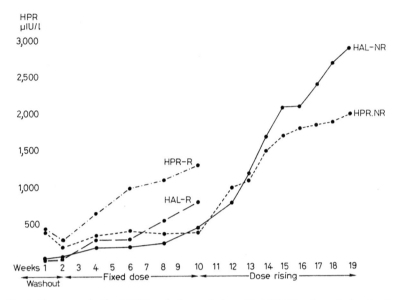

Fig. 1. Plasma prolactin (HPR) and plasma haloperidol (HAL) values during treatment on a fixed low dose, and on increasing higher doses of haloperidol. R = Responders; NR = non-responders.

this dose. When the dose of haloperidol was later raised, there was a progressive rise in plasma prolactin and plasma haloperidol in the initial non-responders. Most of these patients went on to show a satisfactory clinical response on the higher dosage. All showed some improvement, and in some a significant degree of remission was achieved.

While plasma haloperidol levels continued to rise as long as the dosage of medication was increased, plasma prolactin levels tended to level off at values of approximately 2,000 μIU/ml, once high plasma haloperidol values had been achieved.

Discussion

Our results support the suggestion that plasma prolactin values provide a useful guide to neuroleptic therapy. It appears probable that an optimal elevation of plasma prolactin must be attained if the drugs are to assert a therapeutic effect. A previous study of prolactin levels during tranquilization after admission to an acute psychiatric ward indicated that most patients show a 3.2–3.8 increase in plasma prolactin during this period [10]. It is interesting that an increase of the same magnitude was found in our initial responder patients. We suggest that this degree of prolactin elevation is necessary if the drugs are to exert a therapeutic effect. Some patients went on to a considerably higher dose of drug, and considerably higher values of plasma haloperidol were achieved. The plasma prolactin response, however, levelled off to a plateau at about 2,000 μIU/ml. This plateau may have been determined by the availability of dopamine receptors in the tuberoinfundibular region. Once all receptors are blocked further increase in dosage or even an increase in plasma level of drug would have no additional effect on disinhibition of prolactin release.

It is unclear from our present data whether any additional therapeutic advantage was obtained by increasing the dosage *above* the maximal prolactin response, but we hope to elucidate that by further analysis of the data.

References

1 Beumont, P.; Corker, C.S.; Friesen, H.G. et al.: The effects of phenothiazines on the endocrine function. Br. J. Psychiat. *124:* 413–430 (1974).
2 Carlsson, A.: Does dopamine play a role in schizophrenia? Psychol. Med. *7:* 587–597 (1977).
3 Snyder, S.Y.: Catecholamines in the brain as mediators of amphetamine psychosis. Archs gen. Psychiat. *27:* 169–179 (1972).
4 Clemens, J.A.; Smalstig, E.B., and Sawyer, B.D.: Antipsychotic drugs stimulate prolactin release. Psychopharmacologica *40:* 123–127 (1974).
5 Sachar, E.J.; Gruen, P.H.; Altman, N. et al.: Use of neuroendocrine techniques in psychopharmacological research; in Sachar, Hormones, behaviour and psychopathology (Raven Press, New York 1976).
6 Forsman, A. and Ohman, R.: On the pharmacokinetics of haloperidol. Nord. psykiat. Tidsskr. *28:* 441–448 (1974).
7 Lorr, M.; McNair, D.M., and Kleet, C.J.: In-Patient Multidimensional Psychiatric Scale (Consult. Psychol. Press, Palo Alto 1962).
8 Forsman, A.; Martensson, E.; Nyberg, G. et al.: A gaschromatographic method for determining haloperidol. Arch. Pharmacol. *286:* 113–124 (1974).
9 Hwang, P.; Guyda, H.A., and Friesen, H.G.: A radioimmunoassay for human prolactin. Proc. natn. Acad. Sci. USA *68:* 1902–1904 (1971).
10 Meltzer, H.Y. and Fang, V.S.: Serum prolactin levels in schizophrenia – effect of antipsychotic drugs: a preliminary report; in Sachar, Hormones, behaviour and psychopathology (Raven Press, New York 1976).

P.J.V. Beumont, MD, Professor of Psychiatry, Department of Psychiatry, University of Sydney, Sydney 2006 (Australia)

General Discussion on Clinical Hyperprolactinemia: Its Effects and Treatment

Chairmen: *A.G. Frantz and C.J. Eastman*

Molinatti: We report here the results of a study in which the validity of nomifensin (Nom, Alival, Hoechst), an antidepressant which activates dopamine transmission mainly by inhibiting dopamine reuptake in the central nervous system, was assessed as a tool for distinguishing 'functional' from tumorous hyperprolactinemia. The criterion adopted to define PRL 'responsiveness' to Nom was derived on the basis of results gathered in puerperal hyper-PRL (decrease of 30%). Administration of Nom (200 mg orally) did not modify significantly baseline plasma PRL in 41 patients with proven or probable tumors. In contrast to these findings, Nom was effective in lowering plasma PRL levels in 16 of 43 patients with hyper-PRL of uncertain etiology. In the remaining 27 patients, the responses were indistinguishable from those of the tumor groups, but 16 of these patients presented with only equivocal alterations of the sella and in 11 the sella was normal. It is proposed that in the latter 27 subjects unresponsiveness to Nom is an early marker for the presence of a pituitary microadenoma. The results suggest that the Nom test is capable of discriminating between individuals with and without pituitary adenomas.

Judd: I just wonder whether the puerperal group is not the group that responds differently. Perhaps it is not really a good control group for the hyperprolactinemic women, with tumors or otherwise since the estrogen levels are so different. As we learned, the effect of dopamine on prolactin secretion may be increased in the high estrogen situation. I wonder whether what you are actually looking at is a supersensitivity, so-called, in those patients in the puerperal period, which distinguishes them from all the other people with hyperprolactinemia.

L'Hermite: I wonder whether one can really use this nomifensin test to separate the people with functional hyperprolactinemia: most people, including ourselves, who have tried to repeat this test have not succeeded in getting the same results as you. In collaboration with Copinschi, we observed that, in normal male patients, the administration of 2×100 mg nomifensin with a 1-hour interval did not affect prolactin levels in the basal state at all. Right after the study of these basal levels, we induced insulinic hypoglycemia, and there was almost no change in the response, except for a very slight diminution.

Dolman: In considering bromocriptine therapy and when to use it, I was interested in a fact brought up by Hardy: this is the question of treating patients with bromocriptine prior to surgery. He said that, in his experience (unpublished data so far), after bromocriptine for an unspecified length of time, there seems to be a rim or shell of coagulum, or some kind of material, around the adenoma so that he is able to take them out more easily. I would appreciate the comments of anyone with experience in this field and who could help us to decide whether to treat these patients, and for how long, before we send them on to surgery.

Cowden: We have seen 1 patient, who was an acromegalic, who had been treated for 6 months with bromocriptine therapy which had very adequately suppressed his growth hormone levels. He was then required to go for surgery, which at that time was very difficult, because of quite extensive calcification in and around the tumor. We found that a little difficult to interpret, whether it was due to the bromocriptine, or whether it was just due to a very aggressive tumor, because we do know that sometimes these tumors calcify anyway.

del Pozo: I should like to see the histological evidence of fibrosis in the pituitary tissue after prolonged treatment with bromocriptine, because we have followed this issue with electron microscopic examinations in rat material and we have always found normal-appearing cells. Also, we do not see any vascular changes that might lead to necrosis or to ischemia of the tissue and subsequent reduction of the cell mass (Stadelmann and del Pozo, unpublished).

Eastman: I think that we should all agree that we treat women who desire fertility. But in clinical practice, one sees many patients with amenorrhea-galactorrhea or sometimes even asymptomatic hyperprolactinemia. Should these be treated or not? Some of the evidence coming through on the fact that tumors may regress or reduce in size with bromocriptine certainly raises this issue. With the use of bromocriptine, all the speakers have shown almost 100% results, and their doses have been quite low. Many clinicians in the audience would have patients on whom they have used very high doses, and the patient has not responded very well. These were not necessarily patients with large tumors. One point of discussion is what happens to bromocriptine in these patients, do they handle it differently? We have had nothing on the pharmacokinetics of bromocriptine, perhaps I could call upon Isaac to discuss bromocriptine, and how it is handled, and the pharmacokinetics of it.

Isaac: From the studies we have at hand, the important things to say are: that the drug appears to be uniformly well-absorbed, from ^{14}C-labelled studies; that the absorption is fairly complete; that, in studies looking at the relationship between peak plasma levels and single doses, there is an almost linear relationship between the peak plasma level, or the area under the plasma level curve, and the dosage, at least within the range 1–7.5 mg, although at the upper dose level, there is perhaps a suggestion that there may be some saturation of first pass effects, and that the peak plasma level is a little more than one would have expected from extrapolation of the figures for 1, 2.5, and 5 mg. The other thing to mention is that the plasma half-life, determined from unchanged bromocriptine using the radioimmunoassay, is quite short – it is around 3 h, and this rather suggests that there is a dissociation between the plasma half-life and the biological half-life of the drug. The only other thing to add is that most of the elimination seems to be through the bile; there is little unchanged drug eliminated through the urine; there are two or three major metabolites which appear to be inactive.

Thorner: We have looked at the acute effects of a single dose of bromocriptine, and then we looked at bromocriptine levels at 3 and 6 months in 7 hyperprolactinemic patients treated with 2.5 mg three times per day. We found that after a single dose of 2.5 mg bromocriptine the peak levels were around 0.5 ng/ml, and the peak levels were seen at about 3 h. This is in contrast to studies in which a 2.5 mg dose was given to normal subjects on an empty stomach: it was found that the peak levels were seen after 1–1 ½ h, which would support the long held view that taking bromocriptine with food does inhibit the absorption, or slows the absorption. That is presumably why the side effects are less bad if the drug is taken with food.

The other thing that we noted was that some of these patients had prolactin levels that were outside the normal range. The levels had fallen, but not into the normal range. There is a spectrum of response. But when we looked at the bromocriptine levels between those patients, we found that the levels were similar, irrespective of the degree of suppression. Furthermore, when we took a few of these patients and increased the dose from 7.5 to 15 to 30 mg/day, we found that, as expected, the bromocriptine levels rose, but we saw no further suppression of the prolactin levels. I am unconvinced that taking the dose up is always right: I think that you should do it, but if you find that there is no further suppression, then you should reduce it again to 7.5 mg/day.

Ajayi: We observed in our clinic that some of the side effects of bromocriptine can be knocked out if you give a calcium-antagonizing substance like verapamil.

Frantz: It is apparent that efforts to discriminate tumorous from nontumorous causes of hyperprolactinemia will continue, but we must bear in mind that it has been evident from what has been presented that the situation is not at all simple. There is great variability, and we really do not know what the basic pathology is, in many cases. Even when there is an obvious tumor, we do not know the mechanisms of it, we do not know to what extent hyperplasia may or may not be associated in adjoining tissue. The meaning of the tests is, therefore, quite hard to interpret, without very good histological control.

We have not heard very much about the natural history of microadenomas. It would be of interest to know what happens to them; if they are followed with serial prolactin determinations and occasional X-rays, do they go on secreting prolactin at the same rate, do they sometimes regress altogether, how many will enlarge into frank macroadenomas? These are unresolved questions.

The results this morning have, again, confirmed the great utility of bromocriptine as an antihyperprolactinemic agent, and in particular as an antitumor agent. The results of McGregor and his associates are of particular interest, indicating the very high prevalence of tumor shrinkage with bromocriptine. This will serve as an encouragement to many of us to use this agent, perhaps as primary therapy, in patients with pituitary tumors, even comparatively large ones. The question of the extent to which this should be used as a prelude to surgery was alluded to in the discussion. Once again, this needs further work. Is bromocriptine, for example, the propylthiouracil of pituitary tumors? We still do not know.